Oceanography for Meteorologists

Oceanography for Meteorologists

H U Sverdrup

2013

BIOTECH BOOKS
DELHI - 110035

ISBN 81-7622-060-4

Published by : **BIOTECH BOOKS**
1123/74, Tri Nagar,
Delhi - 110 035
Phone: 7109765
e-mail: biotechbooks@yahoo.co.in

Printed at : **Chawla Offset Printers**
New Delhi - 110 052

Preface

Up to the present time, no adequate book has existed from which a meteorologist could readily obtain information as to findings in physical oceanography that have bearing upon problems of the atmosphere. Nor has any text been available that, besides describing the methods used in physical oceanography, also contained a summary of our present knowledge of the current systems of the oceans and of the processes that maintain the currents. It is to fill these needs that this book has been written.

The processes of the sea surface are of the most direct meteorological significance. Consequently, the temperature, the salinity, the currents of the upper layers of the oceans, and the factors that control the existing conditions are the features which have been emphasized. Since these cannot be fully understood without taking the subsurface conditions into account, the deep-water circulation of the oceans has also been discussed. The theories of the large-scale ocean currents and of the wind-driven currents have been included, because in oceanography the application of hydrodynamics is of fundamental importance and because the treatment is similar to that of corresponding problems in the atmosphere. Tides and tidal currents have not been discussed, because the tides have not been shown to be of meteorological significance. Similarly, internal waves in the oceans have not been dealt with, but wind waves have been included.

The theoretical discussion of the dynamics of the ocean currents and the factual information from many ocean areas are as yet incomplete, and therefore it may be premature to generalize. Nevertheless, it has been attempted to overcome difficulties arising from differences in interpretation of incomplete data by placing emphasis on application of the equation of continuity in the description of the ocean circulation.

No references to original papers are given, but several standard works are listed, some of which contain numerous references to meteorological and oceanographic literature.

I am much indebted to my colleagues, Messrs. Johnson and Fleming, for their permission to make use of many parts of our common book, now in press: *The Oceans: Their Physics, Chemistry, and General Biology.* I am greatly obliged to Mr. John A. Fleming, Director of the Department of Terrestrial Magnetism, Carnegie Institution of Washington, for allowing me free use of unpublished data from the last cruise of the *Carnegie*. I am also greatly obliged to Mr. W. C. Jacobs for allowing me to use several figures from his as yet unpublished paper on the exchange of heat and water vapor between the oceans and the atmosphere.

Valuable assistance has been received from Mr. L. Lek, who has carried out a large number of calculations of transport by ocean currents and of net heat received by the oceans in the Northern Hemisphere. Most of the illustrations have been prepared especially for either the larger book or for this one, and credit for them is due Mr. E. C. La Fond. The manuscript has been critically examined by Mr. R. H. Fleming, and parts have been read by Messrs. J. Holmboe, of the University of California at Los Angeles, C.-G. Rossby, of the University of Chicago, and C. L. Utterback, of the University of Washington, all of whom have made helpful suggestions. To all these I extend my sincere thanks. Last, but not least, I wish to acknowledge the help received from my secretary, Miss Ruth Ragan, in correcting the manuscript, in reading proof, and in preparing indices.

H. U. Sverdrup

Scripps Institution of Oceanography
University of California
La Jolla, California

Literature

Most of the following selected books in oceanography and meterology contain extensive bibliographies.

Bjerknes, V., and J. W. Sandström. Dynamic meterology and hydrography. Pt. I. (Carnegie Institution, Publ. no. 88.) Washington. 146 pp., 1910.

Bjerknes, V., J. Bjerknes, H. Solberg, and T. Bergeron. Physikalische Hydrodynamik. Berlin. Julius Springer. 797 pp., 1933.

Brunt, David. Physical and dynamical meteorology. New York. MacMillan, 2nd ed., 428 pp., 1939.

Cornish, Vaughan. Ocean waves and kindred geophysical phenomena. London. Cambridge University Press. 164 pp., 1934.

Dorsey, N. Ernest. Properties of ordinary water-substance. (Amer. Chem. Soc., Monograph Ser. no. 181.) New York. Reinhold Pub. Corp., 673 pp., 1940.

Haurwitz, Bernhard. Dynamic meteorology. New York. McGraw-Hill Book Company, 365 pp., 1941.

Krümmel, Otto. Handbuch der Ozeanographie. Stuttgart. J. Engelhorn. Vol. 1, 526 pp., 1907; vol. 2, 764 pp., 1911.

Marmer, H. A. The sea. New York. D. Appleton-Century Company. 312 pp., 1930.

National Research Council. Physics of the earth. Vol. 5, Oceanography (National Research Council Bulletin no. 85), 581 pp., 1932.

Petterssen, Sverre. Weather analysis and forecasting. New York. McGraw-Hill Book Company. 505 pp., 1941.

Sverdrup, H. U., M. W. Johnson, and R. H. Fleming. The oceans: Their physics, chemistry, and general biology. New York. Prentice-Hall. 1942.

U. S. Weather Bureau. Atlas of climatic charts of the oceans. (Weather Bureau No. 1247), 130 pp., 1938.

Table of Contents

List of Illustrations

LIST OF ILLUSTRATIONS

[illegible] *Illustration*

Chart [illegible] [illegible] Summative [illegible]

Chart [illegible] [illegible] of [illegible] [illegible] [illegible] [illegible] North [illegible]

Chart 4. [illegible] of the [illegible] [illegible] March

CHAPTER I

Introduction

The Heat Budget of the Earth

The importance of the oceans to the circulation of the atmosphere is best understood by considering briefly the heat budgets of the earth as a whole and of the atmosphere. The earth as a whole receives and loses energy by processes of radiation. Radiation is described, in general, as a wave motion which, when absorbed in a medium, is transformed into some other form of energy, such as heat. The wave length of radiation varies within very wide limits, but the radiation that is of importance to the energy balance of the earth all falls within a relatively narrow range. The wave lengths will here be given in the unit μ, which is equal to 10^{-3} mm. In this unit the wave lengths of the visible light fall approximately within 0.38 μ and 0.7 μ, and the radiation that is of importance to the heat budget of the earth falls within the limits 0.15 μ and about 120 μ.

The energy which the earth receives by radiation from the sun is in part reflected from the earth's surface or from clouds and is in part absorbed in the atmosphere or at the earth's surface. Only the absorbed amount is of importance to the heat budget of the earth. The earth loses energy by radiation to space, partly directly from the earth's surface and partly from the atmosphere. On an average, the amounts of incoming and outgoing energy are at balance, because otherwise the temperature of the earth would change, and no evidence exists for measurable changes of this kind. In order to study this balance and to examine the heat budget of the atmosphere, it is necessary to enter briefly upon certain characteristics of the radiation from different bodies.

Black-body Radiation. A body of a given temperature emits radiation of different wave lengths, the amount of radiation from a unit surface depending upon the temperature and the character of the radiating body. According to experience, at every temperature there is an upper limit to the amount of radiation that is emitted. A body which at any given temperature emits the maximum amount of radiation at every wave length is called a "black body." For a black body the total amount of energy radiated per unit time from a unit area equals $\sigma T,^4$

where σ is a constant (Stefan's constant) and where T is the absolute temperature. This relation is known as *Stefan's law.* If radiation is measured in gram calories per square centimeter per minute, the numerical value of σ is 82×10^{-12}. The energy of radiation of different wave lengths, λ, is expressed by *Planck's law,* which can be written

$$\frac{E_\lambda}{T^5} = \frac{c_1(\lambda T)^{-5}}{e^{\frac{c_2}{\lambda T}} - 1} = f(\lambda T),$$

where E_λ is the energy emitted per unit area and unit time within a unit range of wave length and where c_1 and c_2 are constants. The function $f(\lambda T)$ is zero for $\lambda = 0$ and $\lambda = \infty$ and is at a maximum when the value of λT equals 2940. Therefore the wave length of maximum radiation,

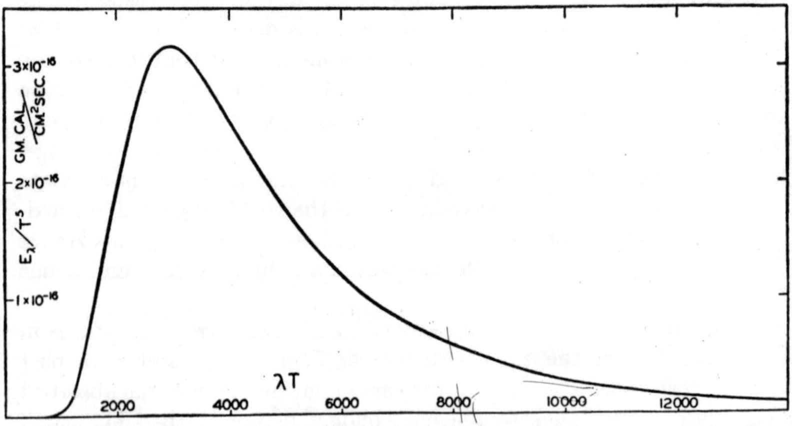

Fig. 1. Energy emitted by a black body at different wave lengths and different absolute temperatures.

λ_m, equals $2940/T$; that is, for bodies of temperature 6000°, 300°, and 200° the maximum energy of the radiation is nearly at wave lengths 0.5 μ, 10 μ, and 15 μ, respectively.

The distribution of the energy can be found by plotting $f(\lambda T)$ against λT (fig. 1). The relation between the emitted energy at a given temperature T and wave length λ is found from this figure by dividing the scale values along the abscissa by T and multiplying the scale values along the ordinate by T^5.

Selective Radiation and Absorption. Most gases and vapors do not radiate as black bodies, but at certain wave lengths they emit radiation of an intensity comparable to the intensity of black-body radiation of the same temperature. If the rate at which a body radiates from unit area is called "emissive power" (Brunt, 1939), it can be stated that the emissive power of a black body is a continuous function of temperature, in accordance with Planck's law, whereas the emissive power of a gas may correspond to that of a black body of the same temperature at a few wave lengths only and may be nearly nil at other wave lengths. The rate at which radiation is absorbed per unit area, the "absorptive power," is related to the emissive power in a manner that is expressed by Kirchhoff's law: *At a given temperature the ratio between the absorptive and emissive power for a given wave length is the same for all bodies.* This law implies that, if a body emits radiation of a given wave length at a given temperature, it will also absorb radiation of that same wave length. Therefore, any gas which emits only a radiation of certain wave lengths also absorbs only radiation of these wave lengths, and this type of radiation and absorption is called *selective.*

RADIATION FROM THE SUN. The radiation from the sun is being examined by the Smithsonian Institution at high-altitude stations in California, Chile, and Sinai. By measurements at high altitudes, with the sun at different distances from zenith, it is possible to draw conclusions as to the solar radiation that reaches the limit of the atmosphere. It has been found that at wave lengths greater than 0.3 μ this radiation has nearly the character of radiation from a black body at an absolute temperature of 5600° to 6000°. The maximum intensity lies close to a wave length of 0.5 μ, and at wave lengths greater than about 2.6 μ the energy is negligible. The radiation from the sun is therefore called *short-wave* radiation. It is continuous except for a few bands that show the effect of absorption in the sun's atmosphere. The intensity of normal incident radiation is, on an average, 1.94 g cal/cm^2/min, and this intensity is called the "solar constant."

Evidence has been accumulated which indicates that in the ultraviolet, at wave lengths shorter than 0.3 μ, the sun does not radiate as a black body but emits a greater and possibly variable amount of energy. This energy is absorbed in the very highest atmosphere and is not measurable. In the following sections it will therefore be necessary to disregard this component of the solar radiation, although it may contribute significantly to the heat balance of the earth and the heat budget of the atmosphere.

Of the incoming radiation a considerable fraction is reflected from the atmosphere, the earth's surface, and the upper surfaces of the clouds. According to Aldrich (Brunt, 1939) the average fraction for the whole earth is 0.43. This figure therefore represents the *albedo* of the earth. The reflected portion of the solar radiation plays no part in the heat budget of the earth. During one year the average solar radiation that reaches the limit of the earth's atmosphere is $S\pi R^2/4\pi R^2 = S/4$, where S is the solar constant and R is the radius of the earth. With $S = 1.94$ g cal/cm^2/min, one obtains $S/4 = 0.485$ g cal/cm^2/min, of which 0.209 is lost by reflection and 0.276 is lost by back radiation to space.

According to Fowle, on a clear day with the sun at zenith, 6 to 8 per cent of solar radiation of wave length greater than 0.3 μ is absorbed in the atmosphere, primarily by the water vapor, which shows some narrow absorption bands in the vicinity of wave length 1.0 μ.

LONG-WAVE RADIATION TO SPACE. The loss of heat by radiation is due in part to radiation from the water vapor in the atmosphere and in part to radiation from the earth's surface, which escapes without being absorbed in the atmosphere. At the prevailing temperature on the earth the maximum intensity of the outgoing radiation lies at wave lengths 10 μ to 15 μ, and this radiation is therefore called *long-wave*.

The water vapor is the only component in the atmosphere which absorbs or emits appreciable amounts of radiation at the temperatures that prevail in the atmosphere, but the radiation and absorption are highly selective. Radiation of certain wave lengths is completely absorbed in a small mass of water vapor, whereas for other wave lengths the water vapor is nearly transparent. Studies of the mechanism of the radiation characteristics of water vapor in the atmosphere are in rapid progress, but a summary of the present status would lie beyond the scope of this brief discussion. It must suffice to state that, owing to the selective absorption of the water vapor, a certain part of the outgoing long-wave radiation comes directly from the earth's surface. This part is of great importance to the heat budget of the earth as a whole, but *does not enter in the heat budget of the atmosphere.*

The Heat Budget of the Atmosphere

Accepting the above values of the solar constant, 1.94 g cal/cm^2/min, and the earth's albedo, 0.43, one obtains that, of the incoming radiation from the sun, an annual average amount of 0.276 g cal/cm^2/min reaches the surface of the earth or is absorbed in the atmosphere. The same average amount must be lost by radiation to space from the water vapor and clouds in the atmosphere or directly from the earth's surface. As yet it is not possible to separate the amounts that are lost by these two processes, but it is possible to state certain limits between which the direct loss from the earth's surface must lie.

The surface of the earth radiates nearly as a black body and emits therefore an amount which approximately equals σT_e^4, where T_e is the temperature of the surface. The earth's surface also receives radiation from the water vapor and the clouds in the atmosphere, but this amount, which may be called R, is nearly always smaller than σT_e^4. The difference, $\sigma T_e^4 - R$, represents the "effective radiation of the earth's surface," or the "nocturnal radiation" (p. 58). The greatest possible loss of radiation directly from the earth's surface would occur if the effective radiation escaped without being absorbed in the atmosphere, and the smallest possible loss would take place if the effective radiation were completely absorbed in the atmosphere. It is probable that the greater part of the effective radiation escapes because the water vapor is nearly transparent for radiation of wave lengths between 8.5 and 11 μ, at which the black-body radiation of the earth's surface is nearly at a maximum.

The average value of the effective radiation over all oceans between 70°N and 70°S is about 0.090 g cal/cm²/min (p. 59). The effective radiation from the land areas between the same latitudes is probably greater, whereas that from the polar regions is smaller. The above value may therefore be taken as a fair estimate of the average value for the earth. Accordingly the radiation from the atmosphere must lie between the limits 0.186 g cal/cm²/min and 0.276 g cal/cm²/min—probably closer to the former. Since the heat gained by the atmosphere by absorption of radiation or by other processes must equal the loss by radiation, the gain must also lie between the same limits.

This gain can be further analyzed, since it may be brought about by (1) condensation of water vapor in the atmosphere, (2) conduction of heat from the earth's surface, (3) absorption of short-wave radiation from the sun, and (4) absorption of long-wave radiation from the earth's surface.

The amount under item (1) could be computed if one knew approximately the total precipitation on the earth. According to Wüst the total precipitation on the whole earth is 396,000 km³/year, corresponding to an average amount of heat released by condensation equal to 0.086 g cal/cm²/min. Under item (2) the average amount of heat conducted from the oceans to the atmosphere between latitudes 70°N and 70°S has been estimated at 0.013 g cal/cm²/min. The conduction from the land areas between the same latitudes is probably smaller, and in the polar region heat is conducted from the atmosphere to the earth's surface. On an average for the whole earth, item (2) can therefore be estimated at 0.010 g cal/cm²/min. The amounts of absorbed radiation, items (3) and (4), have not been estimated, but the limits of the amounts can be found because the sum of all items must lie between 0.186 and 0.276 g cal/cm²/min. On this basis the heat budget of the atmosphere can be summarized as follows:

	Minimum (g cal/cm²/min)	*Maximum* (g cal/cm²/min)
Heat gained by		
Condensation of water vapor	.086	.086
Conduction of heat from the earth's surface	.010	.010
Absorption of radiation	.090	.180
	.186	.276
Heat lost by radiation	.186	.276

It is probable that the true values lie closer to those listed under the minimum heat budget.

All numerical values given here are subject to revision, but such revision is not likely to reverse the feature which the available data bring out—namely, that *the condensation of water vapor in the atmosphere accounts for nearly half the amount of energy which the atmosphere receives.*

The Atmosphere as a Thermodynamic Machine

So far, we have considered only the average heat budget of the atmosphere and have not examined the geographic and seasonal distribution of the regions in which heating and cooling take place. Such an examination cannot be made in detail because our knowledge of the processes of radiation in the atmosphere and of the regions where condensation of water vapor takes place is inadequate. It appears certain, however, that the greater heating takes place in the Tropics and the greater cooling in the high latitudes, so that on an average heat must be transported from the Tropics to the polar regions in order to maintain an approximately stationary distribution of temperature. It is also probable that this transport of heat takes place mainly by air currents, which implies that the atmosphere is a thermodynamic machine that transforms heat into kinetic energy. This kinetic energy is dissipated by friction and transformed back again into heat that is lost by radiation. If the machine runs at a constant speed, the amount of heat which in unit time is transformed into kinetic energy must equal the amount of heat which in unit time is generated by dissipation of kinetic energy.

In a thermodynamic machine that produces kinetic energy the heating must take place, according to a theorem formulated by V. Bjerknes (Bjerknes *et al*, 1933), under a higher pressure and the cooling under a lower pressure. Applied to the atmosphere the theorem states that heating must take place at a lower altitude and cooling at a higher altitude. This condition is fulfilled in the atmosphere because condensation of water vapor does take place at a relatively low altitude; absorption of long-wave radiation from the earth also takes place at relatively low altitudes, whereas radiation to space from the water vapor and the upper surfaces of clouds takes place from high altitudes. Thus, the general features of the thermodynamic machine represented by the atmosphere can be recognized, but the details are not understood. A discussion of the interaction between the oceans and the atmosphere will supply some of the details that have to be considered and is therefore necessary for the better comprehension of the atmospheric circulation.

Several attempts have been made at an examination of the general character of the atmosphere as a thermodynamic machine, the latest and most widely quoted being that by Simpson (Bjerknes *et al*, 1933; Brunt, 1939). Simpson considers only the distributions with latitude of the incoming short-wave radiation and the total outgoing long-wave radiation, making certain simplifying assumptions as to the absorption spectrum of the water vapor. He finds that the earth receives a surplus of radiation up to latitude 35° and concludes that this surplus must be transported by winds to higher latitudes in order to counterbalance the radiation deficit existing there. Simpson does not, however, examine the manner in which the incoming radiation becomes available for heating the atmosphere. He is therefore not concerned with the important part played by condensation of water vapor in the atmosphere. Taking this into account and considering also that part of the outgoing radiation from the earth's surface escapes without influencing the heat budget of the atmosphere, this heat budget becomes much more complicated, and Simpson's numerical values for the heat transport by the atmosphere across parallels of latitude lose much of their significance.

The Oceans as Part of the Subsurface of the Atmosphere

Since the absorption of long-wave radiation from the earth's surface and the condensation of water vapor that is supplied to the atmosphere from the earth's surface are factors of dominant importance to the heat budget of the atmosphere, it is evident that the thermal characteristics of the surface must be examined when studying the atmospheric circulation. Over the land areas, an interaction between the surface and the circulation of the atmosphere evidently takes place. Thus, a covering of snow greatly increases the reflection of incoming short-wave radiation, dense clouds alter the radiation phenomena, heavy rainfalls that moisten the ground change the possibilities for evaporation, and so on, and every change must be reflected in a change in the pattern of atmospheric circulation.

The character of the ocean surface is very different from that of the land, mainly because the ocean water can be set in motion. The heat absorbed near the surface is distributed over a relatively deep water layer through the stirring by waves and wind currents and it is transported over long distances by ocean currents that are more or less directly produced by the prevailing winds. Heat which in one season is absorbed in one part of the sea may therefore in another season be used for evaporation in a different part of the sea, depending upon the types of currents that the general circulation of the atmosphere induces. The oceans, as far as their interaction with the atmosphere is concerned, are a flexible medium that responds to changes in the circulation of the atmosphere. Inasmuch as the greater part of the water vapor in the atmosphere is supplied by evaporation from the oceans, the relation between the atmosphere and the circulation of the oceans becomes important, because this relation will determine the seasons and the regions from which excessive evaporation takes place. Thus the heat budget of the oceans and of the ocean currents is of the greatest importance to the atmosphere. In order to discuss this heat budget and the currents it is necessary first to deal with some of the physical properties of sea water.

CHAPTER II

Physical Properties of Sea Water

The physical properties of pure water depend upon two variables, temperature and pressure (Dorsey, 1940; International Critical Tables), whereas for sea water a third variable has to be considered—namely, the salinity of the water, which will be defined and discussed below. Some of the properties, such as compressibility, thermal expansion, and refractive index, are only slightly altered by the presence of dissolved salts, but other properties that are constant in the case of pure water, such as freezing point and temperature of maximum density, are dependent, in the case of sea water, on salinity. Furthermore, the dissolved salts add new properties, such as osmotic pressure and electrical conductivity.

When dealing with sea water as it occurs in nature, it must also be taken into account that important characteristics are greatly modified by the presence of minute suspended particles or by the state of motion. Thus, the absorption of radiation in the sea is different from the absorption of radiation in pure water or in "pure" sea water because the waters encountered in nature always contain suspended matter that causes increased scattering of the radiation and, consequently, an increased absorption in layers of similar thickness. The processes of heat conduction, chemical diffusion, and transfer of momentum from one layer to another are completely altered in moving water, for which reason the coefficients that have been determined under laboratory conditions must be replaced by corresponding eddy coefficients. Some of the physical properties of sea water, therefore, depend only upon the three variables, temperature, salinity, and pressure, which can all be determined with great accuracy, whereas others depend upon such variables as amount of suspended matter or character of motion, which at present cannot be accurately determined.

Salinity, Temperature, and Pressure

SALINITY. Salinity is commonly defined as the ratio between the weight of the dissolved material and the weight of the sample of sea water, the ratio being stated in parts per thousand or per mille. In oceanography another definition is used because, owing to the complexity of sea water, it is impossible by direct chemical analysis to determine the total quantity of dissolved solids in a given sample. Furthermore, it is impossible to obtain reproducible results by evaporating sea water to dryness and weighing the residue, because certain of the material present, chiefly chloride, is lost in the last stages of drying. These difficulties can be avoided by following a technique yielding reproducible results which, although they do not represent the total quantity of dissolved solids, do represent a quantity of slightly smaller numerical value that is closely related and by definition is called the *salinity* of the water. This technique was established in 1902 by an International Commission, according to which the *salinity* is defined as *the total amount of solid material in grams contained in one kilogram of sea water when all the carbonate has been converted to oxide, the bromine and iodine replaced by chlorine, and all organic matter completely oxidized.*

The determination of salinity by the method of the International Commission is rarely if ever carried out at the present time because it is too difficult and slow, but, inasmuch as it has been established that the dissolved solids are present in constant ratios, the determination of any of the elements occurring in relatively large quantities can be used as a measure of the other elements and of the salinity. Chloride ions make up approximately 55 per cent of the dissolved solids and can be determined with ease and accuracy by titration with silver nitrate, using potassium chromate as indicator. The empirical relationship between salinity and chlorinity, as established by the International Commission, is

$$\text{Salinity} = 0.03 + 1.805 \times \text{Chlorinity}.$$

The chlorinity that appears in this equation is also a *defined* quantity and does not represent the actual amount of chlorine in a sample of sea water. Both salinity and chlorinity are always expressed in grams per kilogram of sea water—that is, in parts per thousand, or per mille, for which the symbol $^\circ/_{\circ\circ}$ is used. In practice, salinity is determined with an accuracy of $\pm 0.02\,^\circ/_{\circ\circ}$.

The salinity of a water sample can also be found by determining the density of the sample at a known temperature, because empirical relationships have been established by which the density can be represented as a function of salinity (or chlorinity) and temperature. Other indirect methods for determining salinity are based on measurements of the electrical conductivity or the refractive index at a known temperature. In order to obtain the needed accuracy, the difference is measured in electrical conductivity or refractive index between the sample and a sample of known salinity that lies fairly close to that of the unknown sample.

The salinity in the oceans is generally between $33\,^\circ/_{\circ\circ}$ and $37\,^\circ/_{\circ\circ}$. In regions of high rainfall or dilution by rivers the surface salinity may be considerably less, and in certain semi-enclosed areas, such as the Gulf of

Bothnia, it may be below 5 °/₀₀. On the other hand, in isolated seas in intermediate latitudes where evaporation is excessive, such as the Red Sea, salinities may reach 40 °/₀₀ or more. As the range in the open oceans is rather small, it is sometimes convenient to use a salinity of 35.00 °/₀₀ as an average for all oceans.

TEMPERATURE. In oceanography the temperature of the water is stated in degrees centigrade. In the oceans the temperature ranges from about $-2°$ to $+30°$C. The lower limit is determined by the formation of ice, and the upper limit is determined by processes of radiation and exchange of heat with the atmosphere (p. 74). In landlocked areas the surface temperature may be higher, but in the open ocean it rarely exceeds 30°C.

PRESSURE. Pressure is measured in units of the c.g.s. system, in which the pressure unit is 1 dyne/cm^2. One million dynes/cm^2 was designated as 1 bar by V. Bjerknes. The corresponding practical unit used in physical oceanography is 1 decibar, which equals $\frac{1}{10}$ bar. The pressure exerted by 1 m of sea water very nearly equals 1 decibar; that is, the hydrostatic pressure in the sea increases by 1 decibar for approximately every meter of depth. *Therefore the depth in meters and the pressure in decibars are expressed by nearly the same numerical value.* This rule is sufficiently accurate when considering the effect of pressure on the physical properties of the water, but details of the pressure distribution must be computed from the density distribution (p. 99).

Owing to the character of the distribution of temperature and salinity in the oceans, some relationships exist between these quantities and the pressure. The temperature of the deep and bottom water of the oceans is always low, varying between 4° and $-1°$C, and high pressures are therefore associated with low temperatures. Similarly, the salinity of deep and bottom water varies within narrow limits, 34.5 °/₀₀ – 35.0 °/₀₀, and high pressures are therefore associated with salinities between these limits. Exceptions are found in isolated seas in intermediate latitudes such as the Mediterranean and Red Seas, where water of high temperature and high salinity is found at great depths—that is, under great pressures.

Density of Sea Water

The density of any substance is defined as the mass per unit volume. Thus, in the c.g.s. system, density is stated in grams per cubic centimeter. The specific gravity is defined as the ratio of the density to that of distilled water at a given temperature and under atmospheric pressure. In the c.g.s. system the density of distilled water at 4°C is equal to unity. In oceanography specific gravities are now always referred to distilled water at 4°C and are therefore numerically identical with densities. The term *density* is generally used, although, strictly speaking, specific gravity is always considered.

The density of sea water depends upon three variables: temperature, salinity, and pressure. These are indicated by using for the density the symbol $\rho_{s,\vartheta,p}$, but, when dealing with numerical values, space is saved by introducing the symbol $\sigma_{s,\vartheta,p}$, which is defined in the following manner:

$$\sigma_{s,\vartheta,p} = (\rho_{s,\vartheta,p} - 1)1000. \qquad \text{(II, 1)}$$

Thus, if $\rho_{s,\vartheta,p} = 1.02575$, $\sigma_{s,\vartheta,p} = 25.75$. The density of a sea-water sample at the temperature and pressure at which it was collected is called the density *in situ* and is generally expressed as $\sigma_{s,\vartheta,p}$. At atmospheric pressure and temperature ϑ°C the corresponding quantity is simply written σ_t, and at 0° it is written σ_0. The symbol ϑ will be used for temperature except when writing σ_t, where, following common practice, t stands for temperature.

At atmospheric pressure and at temperature of ϑ°C the density is a function of the salinity only, or, as a simple relationship exists between salinity and chlorinity, the density can be considered a function of chlorinity. The International Commission, which determined the relation between salinity and chlorinity and developed the standard technique for determinations of chlorinity by titration, also determined the density of sea water at 0° with a high degree of accuracy, using pycnometers. From these determinations the following relation between σ_0 and chlorinity was derived:

$$\sigma_0 = -0.069 + 1.4708\ \text{Cl} - 0.001570\ \text{Cl}^2 + 0.0000398\ \text{Cl}^3.$$

Corresponding values of σ_0, chlorinity, and salinity are given in Knudsen's Hydrographical Tables for each 0.01 per mille Cl.

In order to find the density of sea water at other temperatures and pressures, the effects of thermal expansion and compressibility on the density must be known. The coefficient of thermal expansion has been determined in the laboratory under atmospheric pressure, and according to these determinations the density under atmospheric pressure and at temperature ϑ can be written in the form

$$\sigma_t = \sigma_0 - D. \qquad \text{(II, 2)}$$

The quantity D is expressed as a complicated function of σ_0 and temperature, and is tabulated in Knudsen's Hydrographical Tables. The values of σ_t are widely used in dynamical oceanography. Knudsen's tables also contain a tabulation of D as a function of σ_t and temperature, by means of which σ_0 can be found if σ_t is known ($\sigma_0 = \sigma_t + D$). This table is useful for obtaining the salinity of a water sample the density of which has been directly determined at some known temperature.

The effect on the density of the compressibility of sea water of different salinities and at different temperatures and pressures was examined by Ekman, who established a complicated empirical formula for the *mean* compressibility between pressures 0 and p decibars. From this formula, correction terms have been computed which, added to the value of σ_t, give the corresponding value $\sigma_{s,\vartheta,p}$ for any value of pressure.

COMPUTATION OF DENSITY AND SPECIFIC VOLUME IN SITU. Tables from which the density *in situ*, $\sigma_{s,\vartheta,p}$, could be obtained directly from the temperature, salinity, and pressure with sufficiently close intervals in the three variables would fill many large volumes, but by means of various artifices convenient tables have been prepared. Following the procedure of Bjerknes and Sandström (1910), one can write

$$\rho_{s,\vartheta,p} = \rho_{35,0,0} + \epsilon_s + \epsilon_\vartheta + \epsilon_{s,\vartheta} + \epsilon_p + \epsilon_{s,p} + \epsilon_{\vartheta,p} + \epsilon_{s,\vartheta,p}. \qquad \text{(II, 3)}$$

The first four terms can be expressed by σ_t, which can readily be determined by the methods outlined above, and the remaining terms represent the effects of the compressibility.

Instead of the density *in situ*, its reciprocal value, the specific volume *in situ*, $\alpha_{s,\vartheta,p}$, is generally used in dynamic oceanography. In order to avoid operating with large numbers, the specific volume is commonly expressed as an anomaly, δ, defined in the following way:

$$\delta = \alpha_{s,\vartheta,p} - \alpha_{35,0,p}, \qquad \text{(II, 4)}$$

where $\alpha_{35,0,p}$ is the specific volume of water of salinity, 35°/₀₀ and 0°C, at the pressure p in decibars. The anomaly depends on the temperature, salinity, and pressure, and hence can be expressed as

$$\delta = \delta_s + \delta_\vartheta + \delta_{s,\vartheta} + \delta_{s,p} + \delta_{\vartheta,p} + \delta_{s,\vartheta,p}. \qquad \text{(II, 5)}$$

It should be observed that the anomaly, by definition, does not contain a term δ_p, which would represent the effect of pressure at temperature 0° and salinity 35.00 °/₀₀. The reason for this is explained on p. 100. Of the above terms, the last one, $\delta_{s,\vartheta,p}$, is so small that it can always be neglected. Thus, five terms are needed for obtaining δ, and these were tabulated by Bjerknes and Sandström (1910). If σ_t has already been computed, the terms that are independent of pressure can be combined as $\Delta_{s,\vartheta}$:

$$\Delta_{s,\vartheta} = 0.02736 - \frac{10^{-3}\sigma_t}{1 + 10^{-3}\sigma_t}, \qquad \text{(II, 6)}$$

and the specific volume anomaly can be computed by means of three small tables.

Thermal Properties of Sea Water

THERMAL EXPANSION. The coefficient of thermal expansion, e, of sea water as defined by $e = 1/\alpha\ (\partial\alpha/\partial\vartheta)$ increases with increasing temperature and pressure and is somewhat greater than the corresponding coefficient for pure water. Thus, for sea water of salinity 35.00 °/₀₀ the coefficient at 0° and atmospheric pressure is 51×10^{-6}, and at 20° it is 257×10^{-6}, whereas the corresponding values for pure water are -67×10^{-6} and 207×10^{-6}. The minus sign shows contraction. For sea water of salinity 35.00 °/₀₀ and temperature 0° the coefficient at atmospheric pressure is 51×10^{-6}, and at a pressure of 4000 decibars it is 152×10^{-6}.

THERMAL CONDUCTIVITY. The thermal conductivity is defined as the amount of heat in gram calories which, per square centimeter per second, is conducted through a surface if the temperature gradient at right angles to the surface is one degree centigrade per centimeter. For pure water at 15°C the coefficient equals 1.39×10^{-3}. The coefficient is somewhat smaller for sea water and increases with increasing temperature and pressure. This coefficient is valid, however, only if the water is at rest or in laminar motion (p. 17); but in the oceans the water is nearly always in a state of turbulent motion in which the processes of heat transfer are completely altered. In these circumstances the above coefficient of heat conductivity must be replaced by an "eddy" coefficient which is many times larger and which depends so much upon the state of motion that effects of temperature and pressure can be disregarded.

SPECIFIC HEAT. The specific heat is the number of calories required to increase the temperature of 1 g of a substance 1°C. When studying liquids, the specific heat at constant pressure, c_p, is the property usually measured, but in certain problems the specific heat at constant volume, c_v, must be known.

The specific heat of sea water is somewhat lower than that of pure water and decreases somewhat with increasing temperature. Thus, at atmospheric pressure, the specific heat of sea water of salinity 34.85 °/₀₀ equals 0.941 at a temperature of 0°C, and 0.932 at a temperature of 20°C. The specific heat increases somewhat with increasing pressure. At 4000 decibars it equals 0.970 for water of salinity 34.85 °/₀₀ and temperature 0°C.

The specific heat at constant volume, c_v, is somewhat less than c_p at atmospheric pressure. The ratio c_p/c_v for water of salinity 34.85 °/₀₀ increases from 1.0004 at a temperature of 0° to 1.0207 at 30°. The effect of pressure is appreciable, and for the same water at 0° the ratio is 1.0009 at 1000 decibars and 1.0126 at 10,000 decibars.

LATENT HEAT OF EVAPORATION. The latent heat of evaporation of pure water is defined as the amount of heat in gram calories needed for evaporating 1 g of water, or as the amount of heat needed for producing 1 g of water vapor of the same temperature as the water. Only in the latter form is the definition applicable to sea water. The latent heat of evaporation of sea water has not been examined, but it is generally assumed that the difference between that and pure water is insignificant;

therefore, between temperatures of 0° and 30°C, the formula

$$L_\vartheta = 596 - 0.52\vartheta \qquad \text{(II, 7)}$$

can be used.

Adiabatic Temperature Changes in the Sea. Because sea water is compressible, adiabatic temperature changes take place if a mass of water is brought from one pressure to another without loss or gain of heat. The temperature that a water sample would attain if raised adiabatically to the sea surface has been called the *potential temperature* (Helland-Hansen). Helland-Hansen has prepared convenient tables for computing the potential temperature, which depends upon three vari-

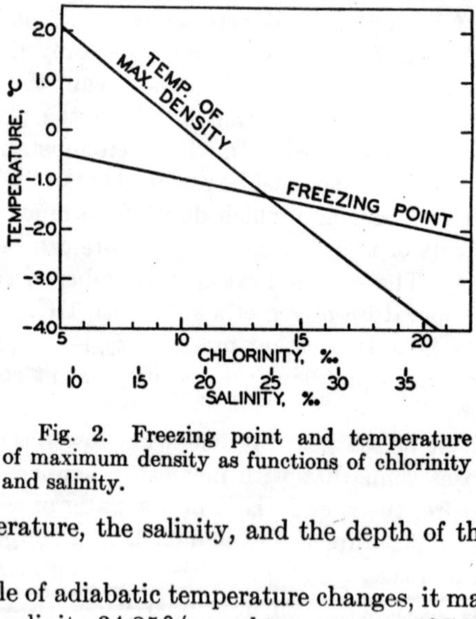

Fig. 2. Freezing point and temperature of maximum density as functions of chlorinity and salinity.

ables: the temperature, the salinity, and the depth of the sample under consideration.

As an example of adiabatic temperature changes, it may be mentioned that, if water of salinity 34.85 ‰ and temperature 2°C is raised adiabatically from a depth of 8000 m to the surface, the temperature drops 0.925°C and the potential temperature of that water is therefore 1.075°C.

In some ocean basins the temperature *in situ* increases toward the bottom, but only in a few isolated deeps is the potential temperature of the water constant.

Freezing-Point Depression and Vapor-Pressure Lowering

Freezing-point depression and vapor-pressure lowering are unique properties of solutions. If the magnitude of one of them is known for a solution under a given set of conditions, the other may be readily computed. Only the depression of the freezing point for sea water of different chlorinities has been determined experimentally, and empirical equations for computing the vapor-pressure lowering have been based on these observations. The freezing point of sea water as a function of salinity and chlorinity is shown in fig. 2.

The freezing point of sea water is the "initial" freezing point—namely, the temperature at which an infinitely small amount of ice is in equilibrium with the solution. If ice forms in a closed system, the concentration of the dissolved solids increases, and hence the formation of additional ice can take place only at lower temperatures.

The vapor pressure of sea water of any salinity S (in per mille) referred to distilled water at the same temperature can be computed from the following equation:

$$e_w = e_d(1 - 0.00053\ \mathrm{S}), \qquad \text{(II, 8)}$$

where e_w is the vapor pressure of the sample and e_d is the vapor pressure of distilled water at the same temperature. The vapor pressure of sea water within the normal range of concentration is about 98 per cent of that of pure water at the same temperature, and in most cases it is not necessary to consider the effect of differences in salinity, since variation in the temperature of the surface waters has a much greater effect upon the vapor pressure.

Other Properties of Sea Water

Maximum Density. Pure water has its maximum density at a temperature of very nearly 4°, but for sea water the temperature of maximum density decreases with increasing salinity, and at salinities greater than 24.70 °/₀₀ it is below the freezing point. At a salinity of 24.70 °/₀₀, the temperature of maximum density coincides with the freezing point: $\vartheta_f = -1.332°$. Consequently, the density of sea water of salinity greater than 24.70 °/₀₀ increases continuously when such water is cooled to its freezing point. The temperature of maximum density is shown in fig. 1 as a function of salinity and chlorinity.

Compressibility. Ekman has derived an empirical equation for the *mean* compressibility of sea water between pressures 0 and p bars, as defined by $\alpha_{s,\vartheta,p} = \alpha_{s,\vartheta,0}(1 - kp)$. The numerical value decreases with increasing temperature and increasing pressure and varies approximately between the limits 4.6×10^{-5} and 4.0×10^{-5}.

The *true* compressibility of sea water is described by means of a coefficient that represents the proportional change in specific volume if the hydrostatic pressure is increased by one unit of pressure: $K = (-1/\alpha)(\partial\alpha/\partial p)$. It can be computed if the mean compressibility is known.

Viscosity. When the velocity of moving water varies in space, frictional stresses are present. The frictional stress, τ, which is exerted on a surface of area 1 cm² is proportional to the change of velocity per centimeter along a line normal to that surface ($\tau = \mu\, dv/dn$), the coefficient of proportionality (μ) being called the *dynamic viscosity*. This coefficient decreases rapidly with increasing temperature and increases slowly with increasing salinity. At a salinity of 35.00 °/₀₀ the viscosity at 0°C is 18.9×10^{-3} g/cm/sec, and at 25°C it is 9.6×10^{-3} g/cm/sec. The effect of pressure is small.

This coefficient of viscosity is valid only if the water is in laminar motion, but, as stated above, since turbulent motion prevails in the sea, an "eddy" coefficient must be introduced which is many times larger (p. 19).

Diffusion. In a solution in which the concentration of a dissolved substance varies in space, the amount of that substance which per second diffuses through a surface of area 1 cm² is proportional to the change

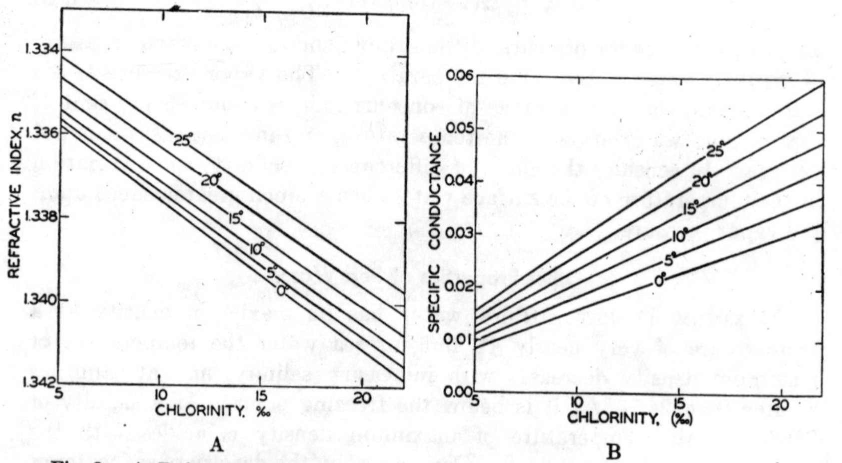

Fig. 3. A. Refractive index of sea water as a function of temperature and salinity. B. Specific conductance, reciprocal ohms per cubic centimeter, of sea water as a function of temperature and chlorinity.

in concentration per centimeter along a line normal to that surface ($dM/dt = -\delta\, dC/dn$). In the c.g.s. system, M is measured in grams per square centimeter, and the concentration C in grams per cubic centimeter. The coefficient of proportionality (δ) is called the *coefficient of diffusion;* for water it is equal to about 2×10^{-5} cm²/sec, depending upon the character of the solute, and is nearly independent of temperature. Within the range of concentrations encountered in the sea the coefficient is also nearly independent of the salinity.

The statement concerning the coefficient of thermal conductivity in the sea applies also to the coefficient of diffusion. Where turbulent motion prevails it is necessary to introduce an "eddy" coefficient that is many times larger and that is mainly dependent on the state of motion.

Surface Tension. The surface tension of sea water is slightly greater than that of pure water at the same temperature.

Surface tension (dynes/cm^2) = 75.64 − 0.144 + 0.0399 Cl ‰. The surface tension is decreased by impurities, and in the sea is mostly smaller than stated.

REFRACTIVE INDEX. The refractive index, n, increases with increasing salinity and decreasing temperatures. It also varies with the wave length of light, and hence a standard must be selected, usually the D line of sodium. In fig. 3A the relation between refractive index, temperature, and chlorinity is shown.

ELECTRIC CONDUCTIVITY. The electric conductivity of sea water depends on temperature and salinity (chlorinity). The specific conductance of sea water is shown in fig. 3B as a function of these two variables.

Eddy Viscosity, Conductivity, and Diffusivity

GENERAL CHARACTER OF EDDY COEFFICIENTS. It has been repeatedly stated that the coefficients of viscosity, heat conduction, and diffusion that have been dealt with so far are applicable only if the water is at rest or in laminar flow. By *laminar flow* is understood a state in which sheets (laminae) of liquids move in such an orderly manner that random fluctuations of velocity do not occur. However, the molecules of the liquid, including those of dissolved substances, move at random, and, owing to this random motion, an exchange of *molecules* takes place between adjacent layers.

In nature laminar flow is rarely or never encountered, but, instead, *turbulent flow*, or *turbulence*, prevails. By *turbulent flow* is understood a state in which random motion of smaller or larger masses of the fluid are superimposed upon some simple pattern of flow. The character of the turbulence depends upon a number of factors, such as the average velocity of the flow, the average velocity gradients, and the boundaries of the system. In the presence of turbulence the exchange between adjacent moving layers is not limited to the interchange of molecules, but masses of different dimensions also pass from one layer to another, carrying with them their characteristic properties. As a consequence the instantaneous distributions of velocity, temperature, salinity, and other variables in the sea are most complicated, and, so far, no means have been developed for studying these distributions. Measurements by sensitive meters have demonstrated that in a given locality the velocity of a current in the sea fluctuates from second to second, as does the wind velocity, but in most cases observations of ocean currents give information only as to *mean* velocities for time intervals that may vary in length from a few minutes to twenty-four hours or more. Similarly, special measurements have demonstrated that the details of the temperature distribution are very complicated, but in general observations are made at such great distances apart that only the major features of the distribution are obtained. Inasmuch as it is impossible to observe the *instantaneous* distributions in space of velocity, temperature, and salinity, it follows that the corresponding gradients cannot be determined and that no basis exists for application to the processes in the sea of the coefficients of viscosity, thermal conductivity, and diffusion that have been determined in the laboratory. Since only certain *average* gradients can be determined, the problem of the effect of turbulence has to be approached in a manner similar to that employed when dealing with the effect of atmospheric turbulence. In order to illustrate this approach, let us first consider the viscosity.

In the case of *laminar* flow in the x-direction only, the *dynamic viscosity*, μ, is defined by the equation $\tau = \mu\, dv/dn$, where τ is the shearing stress and dv/dn is the velocity gradient normal to the surface upon which the stress is exerted. In the case of *turbulent* flow, a *dynamic eddy viscosity*, μ_e, can be defined in a similar manner: $\tau = \mu_e\, d\bar{v}/dn$, where τ now is called the *Reynolds stress* and where $d\bar{v}/dn$ represents the gradient of the observed velocities. The numerical value of the eddy viscosity depends upon the size and intensity of the eddies—that is, on the magnitude of the exchange of masses between adjacent layers. The numerical value of μ_e also depends upon how the "average" velocities have been determined—that is, upon the distribution in space of the observations and upon the length of the time intervals to which the averages refer.

The definition of the eddy viscosity in the above manner appears purely formalistic, but it is based on the concept that masses leaving one layer carry with them the momentum corresponding to the *average* velocity in that layer, and that by impact they attain the momentum corresponding to the average velocity of their new surroundings before again leaving them. Thus, μ_e is an expression for the transfer of momentum of mean motion. This transfer is much increased by the turbulence, as is evident from the fact that the eddy viscosity is many times greater than the molecular viscosity.

In both the atmosphere and the sea it has been found practicable to distinguish between two types of turbulence—vertical and horizontal. In the case of vertical turbulence the *effective* exchange of masses is related to comparatively slight random motion in a vertical direction or, if the term "eddy motion" is used, to small eddies in a vertical plane. Actually, the eddies are oriented at random, but only their vertical components produce any effect on the mean motion. The corresponding eddy viscosity has been found to vary between 1 and 1000 c.g.s. units, thus being one thousand to one million times greater than the molecular viscosity of water. In the case of horizontal turbulence the *effective* exchange of masses is due to the existence of large quasi-horizontal eddies. The corresponding eddy viscosity depends upon the dimensions of the system under consideration and has been found to vary between 10^6 and 10^8 c.g.s. units.

The distinction between vertical and horizontal turbulence is particularly significant where the density of the water increases with depth, because such an increase influences the two types of turbulence in a different manner. Where the density of the sea water increases with depth (disregarding the effect of pressure), vertical random motion is impeded by Archimedean forces, because a mass that is brought to a higher level will be surrounded by water of less density and will tend to sink back to the level from which it came, and, conversely, a water mass moving downward will be surrounded by denser water and will tend to rise. In this case the stratification of the water is called *stable*, because it cannot be altered unless work against gravity is performed. Stable stratification reduces the vertical turbulence, and, where the stability is very great, the vertical turbulence may become nearly suppressed and the eddy viscosity very small. The effect of stability on the horizontal turbulence, however, is probably negligible, because the horizontal random motion takes place mainly along surfaces of equal density. It has even been suggested that the horizontal turbulence increases with increasing stability.

Similar reasoning is applicable to eddy conductivity. In the case of eddy viscosity, it was assumed that the exchange of mass leads to a transfer of momentum from one layer to another, which is expressed by means of μ_e. Correspondingly, when dealing with eddy conductivity, one can assume that the transfer of heat through a unit surface in unit time, dQ/dt, is proportional to the exchange of mass through the surface, as expressed by μ_e, and to the gradient of the potential temperatures, $-d\vartheta/dn$; that is, $dQ/dt = -r\mu_e\, d\vartheta/dn$, where r is a factor that depends upon the specific heat of the fluid and upon the manner in which the heat contents of the moving masses are given off to the surroundings. A mass which is transferred to a new level may break down at that level into smaller and smaller elements, so that equalization of temperature ultimately takes place by molecular heat conduction between the smallest elements and the surroundings. If such is the case, both the difference in momentum and the difference in heat content are given off, and the proportionality factor, r, is *equal to the specific heat of the liquid*. Since the specific heat of water is nearly unity, the numerical values of eddy conductivity and eddy viscosity would be practically equal. However, where stable stratification prevails, the elements, being lighter or heavier than their surroundings, may return to their original level before completion of temperature equalization, whereas equalization of momentum may have been accomplished by collision. In this case, the factor of proportionality, r, will be smaller than the specific heat of the liquid; that is, in the sea, r will be smaller than unity and the eddy conductivity will be smaller than the eddy viscosity. Thus, it may be concluded that stable stratification reduces the vertical eddy conductivity even more than it reduces the vertical eddy viscosity. This conclusion has been confirmed by observation.

The discussion has so far been limited to a consideration of the vertical eddy conductivity, but horizontal eddy conductivity due to horizontal turbulence has also to be introduced. The numerical value of horizontal eddy conductivity must be nearly equal to that of the horizontal eddy viscosity, because the horizontal turbulence is not affected by stable stratification.

The transfer of salinity or other concentration is similar to the heat transfer. The eddy diffusivity is also proportional to the exchange of mass as expressed by μ_e, the factor of proportionality being a pure number. In sea water of uniform density, $r = 1$, but, in the case of stable stratification, when complete equalization of concentration does not take place, $r < 1$; that is, the vertical eddy diffusivity is smaller than μ_e and equals the eddy conductivity. This conclusion has also been confirmed by observation.

INFLUENCE OF STABILITY ON TURBULENCE. The relation between eddy viscosity and eddy diffusion has been examined by Taylor, whose reasoning is based on the fact that in the presence of turbulence the kinetic energy of the system can be considered as composed of two parts: the kinetic energy of the mean motion and the kinetic energy of the superimposed turbulent motion. In homogeneous water the latter is dissipated by viscosity only, and, if the turbulence remains constant, turbulent energy must enter a unit volume at the same rate at which it is dissipated. Where stable stratification is found, part of the turbulent energy is also used for increasing the gravitational potential energy of the system. In this case the rate at which turbulent energy enters a unit volume, T, must equal the sum of the rate at which the potential energy increases, P, and the rate at which energy is dissipated by viscosity, D. It follows that if the turbulence remains unaltered one must have $T > P$.

Taylor shows that the rate at which turbulent energy enters a unit volume equals $\mu_e(dv/dz)^2$, where μ_e is the eddy viscosity, and that the rate at which the potential energy increases equals $gE\,\mu_s$, where E is the stability (p. 100) and where $\mu_s = r\mu_e$ is the eddy diffusivity. It follows that

$$\frac{\left(\frac{dv}{dz}\right)^2}{gE} > \frac{\mu_s}{\mu_e} = r \qquad \text{(II, 9)}$$

is a condition which must be fulfilled if the turbulence shall not be destroyed by viscosity and die off.

Taylor tested the correctness of this conclusion by measurements made by Jacobsen in Danish waters, where Jacobsen found values of μ_e between 1.9 and 3.8 g/cm/sec. Thus, the eddy viscosity was about one hundred to two hundred times greater than the dynamic viscosity of sea water, indicating that turbulence prevailed although the stability reached values as high as $1500 \times 10^{-6}m^{-1}$ (table 1, p. 23). The eddy diffusivity ranged, however, between 0.05 and 0.06 g/cm/sec, and the ratio μ_s/μ_e varied from 0.02 to 0.20, but was in all instances smaller than the ratio $(dv/dz)^2/gE$, as is required by the theory.

According to Taylor's theory, turbulence can *always* be present regardless of how great the stability is, but the type of turbulence must be such that the condition (II, 9) is fulfilled; that is, the rate at which the Reynolds stresses communicate energy to a region must be greater than the rate at which the potential energy of that region increases. This explains why observations in the ocean mostly give smaller values of μ_s than of μ_e. A velocity gradient of 0.1 m/sec on 100 m is common where the stability is about $10^{-6}m^{-1}$, and with these values $\mu_s < 0.1\,\mu_e$. Within layers of very great stability the velocity gradient also is generally great, but the value of μ_s becomes even smaller than in the above example. Thus, below the Equatorial Countercurrent in the Atlantic the decrease of velocity in a vertical direction is $6.10^{-3}sec^{-1}$ and the stability is about $5.10^{-3}m^{-1}$. With these values, $\mu_s < 7 \times 10^{-2}\,\mu_e$, or, if the eddy viscosity were equal to 1 g/cm/sec, the eddy diffusivity would be less than 0.07; that is, it would approach the value of molecular diffusion.

No theory has been developed for the state of turbulence which at indifferent equilibrium or at stable stratification characterizes a given pattern of flow, except in the immediate vicinity of a solid boundary surface (p. 119), but it has been demonstrated that the turbulence as expressed by the eddy viscosity decreases with increasing stability. Thus, Fjeldstad found that he could obtain satisfactory agreement between observed and computed tidal currents in shallow water by assuming that the eddy viscosity was a function not only of the distance from the bottom but of the stability as well. He introduced the equation $\mu_e = f(z)/(1 + aE)$, where E is the stability and where the factor a was determined empirically.

Thus the effect of stability on turbulence is twofold. In the first place the turbulence is reduced, leading to smaller values of the eddy viscosity, and, in the second place, the type of the turbulence is altered in such a manner that the accompanying eddy diffusivity becomes smaller than the eddy viscosity. The latter change is explained by Jacobsen, who assumed that elements in turbulent motion rapidly give off their momentum to their surroundings, but that other properties are exchanged slowly, and that before equalization has taken place the elements are moved to new surroundings by gravitational forces (p. 19). A possible influence of stability on horizontal turbulence has not been examined, but it has been suggested by Parr that this kind of turbulence increases with increasing stability.

Numerical Values of the Vertical and Horizontal Eddy Viscosity and Eddy Conductivity. For the determination of the eddy viscosity, μ_e, in the ocean it is necessary to know the frictional forces and the velocity gradients. The frictional forces cannot be determined directly and must be obtained as the difference between other acting forces, but the velocity gradients can be obtained from directly observed currents. For dealing with wind currents, theories based on certain assumptions as to the character of the eddy viscosity are helpful (p. 124), and for considering currents close to the bottom, results from fluid mechanics permit conclusions as to the eddy viscosity from current measurements alone (p. 119). The uncertainties of the correct application of theories, the difficulties in making current measurements in the open ocean, and the even greater difficulties in determining the acting forces, all introduce great uncertainties in the numerical values of the eddy viscosity that have so far been determined.

Table 1 contains results of such determinations. The specific methods used cannot be discussed here. All values are from regions of moderate or strong currents which have been measured with considerable accuracy or they have been derived from the effect of wind currents on the surface layer. The values range from nearly zero up to 7500, but some consistency in the variations can be seen. At the very bottom, small values are always found, and at greater distances from the bottom great values are associated with strong currents and small stability, whereas small values are associated with weak currents and small stability or with strong currents and great stability. The latter conditions were found on the North Siberian Shelf, where the eddy viscosity approached a small value in a thin layer of very great stability within which the tidal currents reached their maximum values.

The eddy diffusivity, μ_s, can be derived from observed time and space variations of temperature, salinity, oxygen content, or other properties by computations which are based on equations for heat conduction or diffusion. In some cases the currents need not be known, but in other cases only the ratio μ_s/v can be determined, where v is the velocity of the current. If this velocity can be estimated, an approximate value of μ_s can be found. Temperature, salinity, and oxygen content can be measured much more easily and with greater accuracy than currents, and data for studying the eddy diffusion are therefore more readily obtained. As a consequence, many more determinations of eddy diffusion have been made, and the results show greater consistency.

Table 2 contains a summary of values derived from observations in different oceans and at different depths between the surface and the bottom. The values range from 0.02 up to 320, but by far the greater number of values lie between 3 and 90. The range is therefore smaller than the apparent range of the eddy viscosity. The highest value,

TABLE 1

NUMERICAL VALUES OF THE EDDY VISCOSITY, μ_e

Locality	Layer	μ_e in g/cm/sec	μ_e derived from	Authority
All oceans	surface	[a] $\mu_e = 1.02\,W^3$ (W <6m/sec)	Thickness of upper	Thorade, 1914
		$\mu_e = 4.3\ \ W^2$ (W >6m/sec)	homogeneous layer	Ekman, 1905
North Siberian Shelf	0 to 60 m	0–1000	Tidal currents	Sverdrup, 1926
North Siberian Shelf	0 to 60 m	10–400	Tidal currents	Fjeldstad, 1936
North Siberian Shelf	0 to 22 m	[b] $385\left(\frac{z+0.1}{22.1}\right)^{3/4}$	Wind currents	Fjeldstad, 1929
North Sea	0 to 31 m	75–1720	Strong tidal currents	Thorade, 1928
Danish waters	0 to 15 m	[c] 1.9–3.8	All currents	Jacobsen, 1913
Kuroshio	0 to 200 m	[d] 680–7500	All currents	Suda, 1936
Japan Sea	0 to 200 m	150–1460	All currents	Suda, 1936
Off San Diego, Calif	near bottom	[e] $93(z + 0.02)$	Tidal currents	Revelle and Fleming, 1939

[a] W = wind velocity in m/sec.
[b] z = distance from bottom in meters.
[c] Very great stability.
[d] Very strong currents.
[e] z = distance from bottom in meters. **Formula valid between $z = 0.2$ and $z = 1.3$ m.**

TABLE 2

NUMERICAL VALUES OF THE EDDY DIFFUSIVITY AS DERIVED FROM OBSERVED TIME AND SPACE VARIATIONS OF TEMPERATURE, SALINITY, OR OXYGEN CONTENT

Locality	Layer	μ_s in g/cm/sec	Remarks	Authority
Equatorial Atlantic Ocean	0 to 50 m	320	Homogeneous water, moderate currents	Defant, 1932
Bay of Biscay	0 to 100 m	2–16	Moderate stability, weak currents	Fjeldstad, 1933
Danish waters	0 to 15 m	0.02–0.6	Great stability, moderate currents	Jacobsen, 1913
Kuroshio	0 tc 200 m	30–80	Moderate stability, strong currents	Sverdrup *et al*, 1942
Kuroshio	0 to 400 m	7–90	Moderate stability, strong currents	Suda, 1936
Japan Sea	0 to 200 m	1–17	Moderate to small stability, moderate currents	Suda, 1936
California Current	0 to 200 m	30–40	Moderate stability, weak currents	McEwen, 1919
Arctic Ocean	200 to 400 m	20–50	Small stability, weak currents	Sverdrup, 1933
South Atlantic Ocean	400 to 1400 m	5–10	Moderate stability, weak currents	Defant, 1936
Equatorial Atlantic Ocean	600 to 2000 m	8	Moderate stability, weak currents	Seiwell, 1935
Caribbean Sea	500 to 700 m	2.8	Moderate stability, weak currents	Seiwell, 1938
South Atlantic Ocean	3000 m to bottom	4	Homogeneous water, weak currents	Defant, 1936
South Atlantic Ocean	Near bottom	4	Homogeneous water, weak currents	Wattenberg, 1935

TABLE 3

NUMERICAL VALUES OF THE HORIZONTAL EDDY VISCOSITY, $\mu_{e,h}$, AND EDDY DIFFUSIVITY, $\mu_{s,h}$

Locality	Layer	$\mu_{e,h}$ or $\mu_{s,h}$ in g/cm/sec	Remarks	Authority
Northwestern North Atlantic Ocean	Surface	$\mu_{s,h}$ 4×10^8	Strong currents	Neumann, 1940
Atlantic Equatorial Countercurrent	0 to 200 m	$\mu_{e,h}$ 7×10^7	Great stability, moderate currents	Montgomery and Palmén, 1940
	0 to 200 m	$\mu_{s,h}$ 4×10^7		Montgomery, 1939
South Atlantic Ocean	2500 to 4000 m	$\mu_{s,h}$ 10^8	Small stability, weak currents	Sverdrup, 1939
California Current	200 to 400 m	$\mu_{s,h}$ 2×10^6	Moderate stability, weak currents, small area	Sverdrup and Fleming, 1941

320 g cm/sec, was found in a homogeneous surface layer within which one can expect the eddy diffusivity to equal the eddy viscosity. This value is probably the only one that is as great as the corresponding eddy viscosity. Otherwise, the eddy diffusivity is smaller, in agreement with the fact that generally stable stratification is encountered. From the remarks in the table it is evident that the eddy diffusivity increases with increasing velocity of the currents and decreases with increasing stability, in agreement with the preceding considerations.

The results of a few determinations of horizontal eddy coefficients, $\mu_{e,h}$ and $\mu_{s,h}$, are given in table 3. The values of the coefficients have been derived by making bold assumptions and can be considered as indicating only the order of magnitude. This order of magnitude depends upon the size of the area from which observations are available, which probably explains why the value from the California Current is much smaller than the values from the Atlantic Ocean, where conditions within much larger regions have been considered. The latter values, which have been obtained partly from observations at or near the surface and partly from observations at great depth, agree remarkably well. No relation appears between the average velocity of the currents and the eddy coefficients or between the stability and the coefficients. Similarly, from the meager evidence available, the eddy viscosity and eddy diffusion appear to be equal, and therefore the processes of horizontal turbulence seem to be entirely different from those of vertical turbulence, but it should be borne in mind that the study of horizontal turbulence in the ocean is at its very beginning.

Absorption of Radiation

Absorption Coefficients of Distilled Water and of Pure Sea Water. In water the intensity of parallel beams of radiation of wave length λ decreases in the direction of the beams, the decrease in a layer of infinitesimal thickness being proportional to the energy, I_λ, and to the thickness of the layer:

$$dI_\lambda = -\kappa_\lambda I_\lambda \, dz. \qquad \text{(II, 10)}$$

The coefficient of proportionality, κ_λ, is called the absorption coefficient at the wave length λ. Integrating equation (II, 10) between the limits $x = h$ and $x = h + L$, one obtains

$$\kappa_\lambda = \frac{2.30}{L}[\log I_{\lambda,h} - \log I_{\lambda,h+L}], \qquad \text{(II, 11)}$$

where the factor 2.30 enters because base-10 logarithms are used instead of natural logarithms, and where L is the thickness of the layer within which the energy of the radiation is reduced from $I_{\lambda,h}$ to $I_{\lambda,h+L}$. The latter equation also serves as a definition of the absorption coefficient.

The numerical value of the absorption coefficient depends upon the unit of length in which L is expressed. In physics the unit is 1 cm, but in oceanography it has become common practice to use 1 m as the unit of length. Therefore, the numerical values of the coefficients that will be given here are one hundred times larger than the corresponding values given in textbooks of physics.

The decrease of intensity of radiation passing through a layer of water depends not only upon the amount of radiation which is truly absorbed—that is, converted into another form of energy—but also upon the amount which is scattered laterally. In "pure" water the scattering takes place against the water molecules, and the effect of scattering is related to the molecular structure of the water. However, when measuring the absorption in pure water, the effect of scattering is not separated but is included in the absorption coefficient, which varies greatly with wave length.

Extinction Coefficients in the Sea. In oceanography the greater interest is attached to the rate at which downward-traveling radiation decreases. The rate of decrease can be defined by means of a coefficient similar to the absorption coefficient:

$$\kappa_\lambda = 2.30\,(\log I_{\lambda,z} - \log I_{\lambda,z+1}), \qquad \text{(II, 12)}$$

where $I_{\lambda,z}$ and $I_{\lambda,z+1}$ represent the radiation intensities of wave length λ on horizontal surfaces at the depths z and $(z + 1)$ meters. Different names have been proposed for this coefficient, such as "transmissive exponent" or "extinction coefficient." The latter name has been widely used and will be employed here, although the process by which the intensity of radiation is reduced will be called *absorption*. The absorption of radiation in the sea is complicated by the increased scattering due to suspended particles and by the presence of dissolved colored substances. The extinction coefficient of radiation of a given wave length therefore varies within wide limits from one locality to another, and in a given locality it varies with depth and time.

The first crude measurements of absorption in the visible part of the spectrum were made by lowering a white disk of standard size (diameter, 30 cm), the Secchi disk, and observing the depth at which it disappeared from sight. Comparisons with recent exact measurements have shown that the extinction coefficient of visible rays can roughly be obtained from the formula $\kappa = 1.7/D$, where D is the maximum depth of visibility in meters, as determined by the Secchi disk.

The next step in the investigation of the absorption of radiation in sea water was made by subsurface exposure of photographic plates enclosed in watertight containers. Such experiments, which were conducted by Helland-Hansen on the *Michael Sars* Expedition by exposing panchromatic plates at different depths in the vicinity of the Azores, showed that photographic plates were blackened at a very great depth. A plate exposed for 40 minutes at a depth of 500 m showed strong blackening, another exposed for 80 minutes at 1000 m was also blackened, but a third plate which was exposed for 120 minutes at 1700 m showed no effect whatever. These experiments were made at noon on June 6, 1910, with a clear sky. At 500 m it was found that the radiation had a distinct downward direction, because plates exposed at the top of a cube were much more strongly blackened than those exposed on the sides.

In other experiments colored filters were used which showed that the red portion of the spectrum was rapidly absorbed, whereas the green and blue rays penetrated to much greater depths. Quantitative results as to the absorption at different wave lengths were obtained by using spectrophotometers, but the methods were laborious and not sensitive enough to be used to great depths.

The introduction in recent years of photoelectric cells has made possible rapid and accurate determinations of extinction coefficients in various parts of the spectrum. A number of different instruments have been and still are in use, but a standardized technique has been proposed by a committee of the International Council for the Exploration of the Sea. Because of the wide variation in absorption at different wave lengths, efforts have been directed toward measuring exactly the absorption in narrow spectral bands. The determinations are accomplished by lowering stepwise a photoelectric or photronic cell enclosed in a watertight container and provided with suitable color filters, and by observing on deck the photoelectric current by means of a sensitive galvanometer or a suitable bridge circuit. The measurements must be made at constant incident light either on clear, sunny days or on days when the sky is uniformly overcast, because the rapid variations in incident light that occur on days with scattered clouds will naturally lead to erroneous results as to the absorption. If one wishes to determine the percentage amount of radiation that reaches a certain depth, it is necessary to make simultaneous readings of the incident radiation on board ship.

These methods give information as to the absorption in layers of definite thickness. Instruments for measurements of the transparency of sea water at given depths and of the scattering of light have been designed by H. Pettersson and have been used for determining relative values. It has been demonstrated, particularly, that at boundary surfaces sharp variations in transparency and scattering occur. Since the study of the absorption of radiation in the sea is in rapid progress, several of the following generalizations are presented with reservations.

The main results as to the character of the extinction coefficient of radiation of different wave lengths in the sea can be well illustrated by means of data that Utterback, and Jorgensen and Utterback have published. Utterback has attempted to determine the extinction coefficient within as narrow spectral bands as possible and has assigned the observed coefficients to distinct wave lengths, but it should be understood that the wave length actually stands for a spectral band of definite width. He has made numerous observations in the shallow waters near islands in the inner part of Juan de Fuca Strait and at four stations in the open oceanic waters off the coast of Washington, and these can be considered typical of coastal and of oceanic water, respectively. Table 4 contains the absorption coefficients of pure water (according to Sawyer) at the wave lengths used by Utterback, the minimum, average, and maximum extinction coefficients observed in oceanic water, and the minimum, average, and maximum coefficients observed in coastal water. The minimum and maximum coefficients have all been computed from the four lowest and the four highest values in each group. In the clearest oceanic water the extinction coefficients were only twice those of pure water, and the average values were four to five times the latter, whereas the maximum values were up to ten times as great. In the coastal waters the minimum values were up to sixteen times greater than the absorption coefficients of pure water, the average values were up to twenty-four times as great, and the maximum values were up to thirty-four times as great. The increase of the extinction coefficients, however, varied widely in the different parts of the spectrum and was relatively much greater for shorter wave lengths than for longer ones.

TABLE 4

ABSORPTION COEFFICIENTS PER METER IN PURE WATER AND EXTINCTION COEFFICIENTS IN THE SEA
(From Utterback's data)

Type of water	Wave length in μ								
	.46	.48	.515	.53	.565	.60	.66	.80	1.00
Pure water	.015	.015	.018	.021	.033	.125	.280	2 40	39.7
Oceanic water, lowest	.038	.026	.035	.038	.074	.199			
Oceanic water, average	.086	.076	.078	.084	.108	.272			
Oceanic water, highest	.160	.154	.143	.140	.167	.333			
Coastal water, lowest	.224	.230	.192	.169		.375	.477		
Coastal water, average	.362	.334	.276	.269		.437	.623		
Coastal water, highest	.510	.454	.398	.348		.489	.760		

The great difference between the mean and the maximum values of the extinction coefficients shows that the absorption of sea water varies within very wide limits. In the example presented in table 4 the percentage variations are about the same in coastal and oceanic waters, and the maximum values in the oceanic water approach the minimum values in the coastal water. In any given locality the great variations that also occur in a vertical direction further complicate the actual conditions.

Similar results have been obtained by other investigators from such widely different areas as the English Channel, the waters off the east coast of the United States, and those off southern California. In all instances it has been found that the absorption is less in oceanic than in coastal water, but varies within wide limits both locally and with depth. Where examination of absorption in different parts of the spectrum has been conducted, it has been found that the absorption is much less in the blue than in the red end of the spectrum, and that the blue light penetrates to the greatest depths in clear water, whereas the green or yellow light reaches farther down in turbid water.

INFLUENCE OF THE ALTITUDE OF THE SUN UPON THE EXTINCTION COEFFICIENTS. The extinction coefficient is a measure of the reduction of intensity in a vertical distance and therefore depends upon the obliquity of the rays. The obliquity of the incident rays is reduced, however, by refraction when they are entering the water from the air and by the effect of scattering. When the sun's rays pass the water surface, the angle of refraction increases from zero with the sun at zenith to 48.5 degrees with the sun at the horizon, and the most oblique rays penetrating into the water therefore form an angle of less than 48 degrees with the vertical. Owing to the scattering and the sifting out by absorption of the most oblique rays, the measured extinction coefficients will be independent within wide limits of the altitude of the sun. The reduction of the obliquity of the incident radiation has been directly demonstrated by Johnson and Liljequist. Conditions at very low sun have not been examined, but it is probable that the extinction coefficients are then increased, and this may have bearing upon the diurnal variation of the incoming energy at greater depths.

CAUSE OF THE LARGE EXTINCTION COEFFICIENTS IN THE SEA. The fact that extinction coefficients in the sea are large in comparison to those of absolutely pure water is as a rule ascribed to the presence of minute particles which cause scattering and reflection of the radiation and which themselves absorb radiation. According to Lord Rayleigh, if such particles are small in comparison to the wave length, λ, of the radiation, the scattering will be proportional to λ^{-4}, and the effect therefore at wave length, say, .46 will be 2.86 times greater than at wave length .60 μ. This selective effect leads to a shift toward longer wave lengths of the region of minimum absorption.

Clarke and James found that the increased absorption in oceanic water was chiefly caused by suspensoids which could be removed by means of a "fine" Berkefeld filter and that these suspensoids were largely nonselective in their effect. Utterback's data indicated, on the other hand, that the increased absorption in oceanic water is due at least in part to selective scattering, because at short wave lengths the increase of the extinction coefficients was greater than it was at longer wave lengths (table 4). Kalle is of the opinion that selective scattering is of dominant importance, but the question is not yet settled as to the mechanism which leads to increased absorption in oceanic water as compared to pure water. The fact that even in the clearest oceanic water the absorption is greater than in pure water indicates, however, that finely suspended matter is always present. One could state that the ocean waters always contain dust.

The increase of the extinction coefficients in coastal waters appears to be due in part to another process. Clarke and James conclude from their examination that in coastal water both suspensoid and "filter-passing" materials are effective in increasing the absorption, and that each exerts a highly selective action, with greatest absorption at the shorter wave lengths. These great absorptions at the shorter wave lengths are demonstrated by Utterback's measurements (table 4). Clarke does not discuss the nature of the "filter-passing" material, but Kalle has shown that, in sea water, water-soluble pigments of yellow color are present. These pigments appear to be related to the humic acids, but, since their chemical composition has not been thoroughly examined, Kalle calls them "yellow substances." They seem to occur in greatest abundance in coastal areas, but Kalle has demonstrated their presence in the open ocean as well and believes that they represent a fairly stable metabolic product related to the phytoplankton of the sea. The selective absorption of these yellow substances may then be responsible, in part, for the character of the absorption in coastal water and for the shift of the band of minimum absorption toward longer wave lengths.

It has not been possible anywhere to demonstrate any direct influence of phytoplankton populations on the absorption, although it is probable that very dense populations cut down the transparency. At present it appears that the major increase of absorption of sea water over that of pure water is due to two factors: the presence of minute suspended particles, and the presence of dissolved "yellow substances." The latter factor is particularly important in coastal waters.

The Color of Sea Water

The color of sea water as it appears to an observer ashore or on board a vessel varies from a deep blue to an intense green, and is in certain circumstances brown or brown-red. The blue waters are typical of the open oceans, particularly in middle and lower latitudes, whereas the green water is more common in coastal areas, and the brown or "red" water is typical of coastal waters only.

The color of sea water has been examined by observing the color that the water appears to have when seen against the white, submerged surface of the Secchi disk (see p. 27). This color is recorded according to a specially prepared color scale, the "Forell scale." The method is a rough one, and the scale is not adapted to the extreme colors in coastal waters.

Kalle has critically reviewed earlier theories as to the causes of the color of the sea water and arrives at conclusions that appear to be consistent with all available observations. The blue color is explained, in agreement with earlier theories, as a result of a scattering against the water molecules themselves, or against suspended, minute particles smaller than the shortest wave lengths. The blue color of the water is therefore comparable to the blue color of the sky. The transition from blue to green cannot be explained, however, as a result of scattering, and Kalle concludes that it is due to the presence of the water-soluble "yellow substances." He points out that the combination of the yellow color and the "natural" blue of the water leads to a scale of green colors as observed at sea. Fluorescence may contribute to the coloring but appears to be of minor importance.

Suspended larger particles can give color to the sea water if they are present in large quantities. In this case the color is not determined by the optical properties of the water or by dissolved matter, but by the colors of the suspended inorganic or organic particles, and the water is appropriately called "discolored." Discoloring can be observed when large quantities of finely suspended mineral particles are carried into the sea after heavy rainfall or when very large populations, several million cells per liter, of certain species of diatoms or dinoflagellates are present very near the surface. Thus, the "red water" (often more brown than red) which is quite frequently observed in many areas and after which the Red Sea and the Vermilion Sea (the Gulf of California) have been named, is due to an abundance of certain dinoflagellates. Discoloring, however, is a phenomenon of the typical coastal waters, the green colors being frequent in waters near the coast or at sea in high latitudes and the blues characteristic of the open ocean in middle and lower latitudes.

Sea Ice

Freezing and Melting of Ice. If sea water of uniform salinity higher than 25.0 °/₀₀ is subjected to cooling at the surface, the density of the very surface layer increases, giving rise to convection movements that continue until the surface water is cooled to the freezing point, when ice begins to form. At first elongated crystals of pure ice are produced. Since the salinity of the very surface water increases correspondingly, the convective movements are maintained. As the freezing continues, the ice crystals grow into a matrix in which a certain amount of sea water becomes mechanically trapped, and the more rapid the freezing the greater is the amount of sea water enclosed in the ice. If the temperature of the ice thus formed is lowered, part of the trapped sea water freezes, so that the cells containing the liquid brine become smaller and the salt concentration in the enclosed brine becomes greater. Thus, sea ice consists of crystals of pure ice separating numerous small cells containing brine, the concentration of which depends upon the temperature of the ice. If the ice is cooled to very low temperatures, solid salts may crystallize out.

If the temperature of such sea ice rises, the ice surrounding the brine-filled cells melts and separated salt crystals dissolve. As the melting goes on, the brine cells grow in size, and, when the temperature approaches zero, the cells join, permitting the trapped sea water to trickle down. Where the sea ice has been hummocked, all brine will flow down, leaving only the pure ice, which can be used as a source of potable water. In the Arctic, part of the hummocks melt in summer, and the water that then collects in pools on the ice floes is fit for drinking and cooking purposes. Sea ice which has not been exposed to hummocking, on the other hand, will become soggy and will disintegrate.

The salinity of the ice depends upon the rapidity of freezing and upon the temperature changes to which the ice has been subjected. At very rapid freezing, brine and salt crystals may accumulate on ice surfaces, making the surface "wet" at temperatures of $-30°$ to $-40°C$ and greatly increasing the friction against sled runners and skis.

PROPERTIES OF SEA ICE. The properties of sea ice differ greatly from those of fresh-water ice, since they depend upon the amount of enclosed brine and upon the number of air bubbles left in the ice if all or part of the brine has trickled down.

At zero degrees the *density* of pure ice is 0.9168, but the density of sea ice may be both above and below that of pure ice, depending upon its content of brine and air bubbles. Values between 0.92 and 0.86 have been reported. The *specific heat* of pure ice is about half that of pure water. It depends upon the temperature of the ice, but varies within narrow limits. The specific heat of sea ice depends on the temperature and the salinity of the ice, because every lowering of the temperature will involve freezing of some of the enclosed brine, and every raising of the temperature will involve melting of ice surrounding the brine-filled cells. The amounts of heat involved in these processes of freezing and melting are so great that the specific heat is 6.7 at a temperature of $-2°$, 1.99 at $-4°$, 0.88 at $-8°$, and 0.60 at $-16°$. At very low temperatures, when most of the salts have crystallized out, the specific heat approaches that of pure ice.

The *latent heat of fusion* of pure ice at atmospheric pressure is 79.67 g cal/g. No specific value of the heat of fusion can be assigned to sea ice, because, owing to the presence of salts, melting takes place whenever the temperature rises, no matter how low it may be. However, the amount of heat needed for melting sea ice of a given temperature and salinity can be stated. Thus, 68 g cal are needed for melting sea ice of a temperature of $-2°$ and a salinity of 6°/₀₀, and 55 g cal are needed for melting sea ice of a temperature of $-1°$ and the same salinity.

The *latent heat of evaporation* of sea ice probably equals that of pure ice and depends upon whether the ice volatilizes directly to vapor or melts first. In the former case the latent heat of evaporation is about 600 g cal/g, and in the latter case it is about 700 g cal/g. The *vapor pressure* over sea ice cannot depart much from that over pure ice, which is given in meteorological tables.

Pure ice contracts when cooled, but sea ice may expand, depending upon its temperature and salinity, because, when the temperature is lowered, more ice is formed, and because the transformation of brine into ice is accompanied by expansion. Thus, for sea ice of salinity 6°/₀₀ the *coefficient of thermal expansion* at $-2°$ is equal to -69.67×10^{-4}, at $-12°$ it equals 0.00, and at $-22°$ it equals 0.93×10^{-4}. The negative sign indicates that the ice expands when the temperature is lowered.

The coefficient of *thermal conductivity* of pure ice is about 5×10^{-3}. For sea ice the thermal conductivity is in general decreased, owing to the presence of air bubbles in the ice. Values ranging between 1.5×10^{-3} and 5×10^{-3} have been recorded.

The *absorption of radiation* in sea ice has not been examined, but from general experience it can be stated that sea ice has a high transparency for visible radiation. The *temperature radiation* of sea ice nearly equals that of a black body.

The *albedo* of sea ice—that is, the fraction of incoming short-wave radiation which is reflected from the surface of the ice—varies with the character of the surface. For ordinary grayish sea ice the albedo is about 0.50, but for sea ice covered by rime it may be as high as 0.80.

CHAPTER III

Observations In Physical Oceanography

Oceanographic Expeditions and Vessels

The collection of oceanographic data is expensive because it has to be made from vessels whose operation and maintenance are costly. Early oceanographic work was therefore conducted on especially equipped expeditions, most famous among which is the British *Challenger* Expedition of 1873–1876, the first world-wide deep-sea expedition. In the decades following the voyage of the *Challenger*, numerous other expeditions have visited most regions of the oceans, but, in the discussion of the water masses and currents of the oceans, use will primarily be made of data from the more recent ones (listed in table 5) because their data are more accurate. It has become customary to refer to most of the oceanographic expeditions and to their observations and results by the name of the vessel. Thus, the terms *Challenger* or *Atlantis* cruises are employed. Among recent expeditions only the John Murray Expedition to the Indian Ocean is generally referred to by its name and not by the name of the vessel, the *Mabahis*.

In more recent years, deep-sea oceanographic work has also been conducted by government organizations in different countries (hydrographic offices and bureaus of fisheries) and by institutions for oceanographic research. In the United States such work has been carried out by the Hydrographic Office of the U. S. Navy (U.S.S. *Bushnell, Hannibal, Louisville*, and others), the U. S. Coast Guard (*Marion* and others), the U. S. Coast and Geodetic Survey (*Explorer*, *Oceanographer*, and others), and by the Bingham Oceanographic Laboratories, Yale University, the Oceanographic Laboratories of the University of Washington (*Catalyst*), the Scripps Institution of Oceanography, University of California (*E. W. Scripps*), and the Woods Hole Oceanographic Institution (*Atlantis*).

The present technique of making deep-sea observations and the precision instruments used were developed around 1900, when the International Council for the Exploration of the Sea was organized, with headquarters in Copenhagen. At that time intensive exploration of the waters adjacent to northwestern Europe commenced because of the important fisheries there. Subsequently, a number of oceanographic stations have been established, some of which are directly engaged in oceanographic problems related to fisheries, whereas others are independent research institutions. Large vessels such as those used for world-wide expeditions would be far too expensive to operate continuously from such stations, but Helland-Hansen, of the Geophysical Institute in Bergen, Norway, was convinced that small vessels could be used effectively, and in 1913 he had the 76-foot *Armauer Hansen* built to conform to his idea. From this small but sturdy craft, both intensive and extensive oceanographic work has been carried out in the North Atlantic, and the vessel has in every respect answered expectations. Following this lead, other establishments have purchased or built small vessels that can be economically operated.

Every vessel which is fitted out for deep-sea oceanographic work must have adequate winches that are provided with wire ropes for lower-

TABLE 5
SOME OF THE MORE IMPORTANT OCEANOGRAPHIC DEEP-SEA EXPEDITIONS SINCE 1910

Vessel	Nationality	Operations	
		Years	Areas
Armauer Hansen	Norway	1913–	North Atlantic
Atlantis	U.S.A.	1931–	North Atlantic
Carnegie	U.S.A.	1928–1929	North Atlantic, Pacific
Dana	Denmark	1921–1922	North Atlantic
		1928–1930	All oceans
Deutschland	Germany	1911–1912	Atlantic, Antarctic
Discovery	Great Britain	1925–1927	Antarctic, Atlantic
Discovery II	Great Britain	1929–1939	Antarctic, Indian, Atlantic
Mabahis (John Murray Expedition)	Great Britain	1933–1934	Indian Ocean
Meteor	Germany	1925–1927	South Atlantic
		1929–1939	North Atlantic
Michael Sars	Norway	1910	North Atlantic
Willebrord Snellius	Netherlands	1929–1930	Indian Ocean, East Indian Archipelago
William Scoresby	Great Britain	1926–1931	Antarctic, South Pacific
Japanese vessels: *Shintaku Maru*, *Siunpu Maru*, *Soyo Maru*, and others	Japan	1920–	Western North Pacific
U. S. Coast Guard vessels: *Chelan*, *General Greene*, *Marion*, *Oglalla* and others	U.S.A.	1925–	Western North Atlantic, Northern North Pacific

ing instruments to great depths, and it must have laboratory space for such examinations of water samples as need to be made shortly after collection. The arrangements on board ship vary widely, however, depending upon the design of the vessel.

Because of the manner in which oceanographic data have to be obtained, studies in "synoptic oceanography" are in general prohibitive as to cost, and simultaneous observations have not been made except in a few instances when vessels have cooperated within a very limited area. Fortunately, the oceans are much closer to a steady state than is the atmosphere, and therefore observations that have been made by the same ship within a reasonably short time can be treated as if they were simultaneous (p. 106). The now classical example is the work of the International Ice Patrol, which is conducted by the U. S. Coast Guard. During the spring months a vessel of the Ice Patrol makes cruises at intervals of about four weeks, each cruise lasting about two weeks and covering the region to the east and southeast of the Grand Banks of Newfoundland. At the end of each cruise the currents are computed (p. 106), and on the basis of these computations the drift of icebergs is predicted.

Temperature Observations

Three types of temperature-measuring devices are used in oceanographic work: (1) accurate thermometers of standard type are employed for measuring the surface temperature when a sample of the surface water is taken with a bucket, (2) reversing thermometers are used for measuring temperatures at subsurface levels, and (3) thermographs are employed at shore stations and on board vessels for recording the temperature at some fixed level at or near the sea surface. The centigrade scale is the standard for the scientific investigation of the sea. A high degree of accuracy is necessary in temperature measurements because of the relatively large effects that temperature has upon the density and other physical properties and because of the extremely small variations in temperature found at great depths. Subsurface temperatures must be accurate to within less than 0.05°C, and under certain circumstances to within 0.01°C. Such accuracy can be obtained only with well-made thermometers which have been carefully calibrated and rechecked from time to time. Because of the greater variability of conditions in the surface layers, the standards of accuracy there need not be quite so high.

Surface Thermometers. Conventional-type thermometers used for surface temperatures must have an open scale that is easy to read and that is provided with divisions for every 0.1°. The scale should preferably be etched upon the glass of the capillary. The thermometer should be of small thermal capacity in order to attain equilibrium rapidly, it should be checked for calibration errors at a number of points on the scale by comparison with a thermometer of known accuracy, and it should be read with the scale immersed to the height of the mercury column. Observations of surface temperatures made upon bucket samples should be obtained immediately after the sample is taken, because heating or cooling of the water sample by radiation, conduction, and evaporation may have a measurable effect upon the temperature. From a vessel, samples must be taken as far away as possible from any discharge outlets from the hull, and, if the vessel is under way, they should be taken near the bow so as to avoid the churned-up water of the wake.

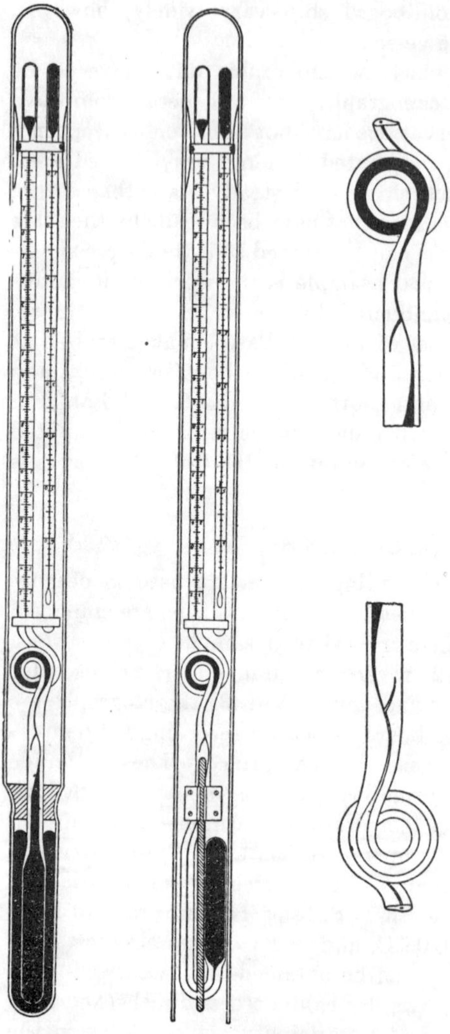

Fig. 4. Protected and unprotected reversing thermometers in set position—that is, before reversal. To the right is shown the constricted part of the capillary in set and reversed positions.

Protected Reversing Thermometers. Reversing thermometers (fig. 4) are usually mounted upon the water sampling bottles (fig. 5), but they may be mounted in reversing frames and used independently. Reversing thermometers were first introduced by Negretti and Zambra (London) in 1874, and since that time have been improved so that at present well-made instruments are accurate to within 0.01°C. On the *Challenger* Expedition the subsurface temperatures were measured by means of minimum thermometers.

A reversing thermometer is essentially a double-ended thermometer. It is lowered to the required depth in the *set* position (fig. 4), and in this position it consists of a large reservoir of mercury connected by means of a fine capillary to a smaller bulb at the upper end. Just above the large reservoir the capillary is constricted and branched, having a small closed arm, and above this the thermometer tube is bent in a loop, from which it continues straight and terminates in the smaller bulb. The thermometer is so constructed that in the *set* position the mercury fills the reservoir, the capillary, and part of the bulb. The amount of mercury above the constriction depends upon the temperature, and, when the thermometer is reversed, by turning through 180 degrees, the mercury column breaks at the point of constriction and runs down, filling the bulb and part of the graduated capillary, and thus indicating the temperature at reversal. The loop in the capillary, which is generally of enlarged diameter, is designed to trap any mercury that is forced past the constriction if the temperature is raised after the thermometer has been reversed. In order to correct the reading for the changes that result from differences between the temperature at reversal and the surrounding temperature at the time of reading, a small standard thermometer known as the *auxiliary thermometer* is mounted alongside the reversing thermometer. The reversing thermometer and the auxiliary thermometer are enclosed in a heavy glass tube that is evacuated except for the portion surrounding the reservoir of the reversing thermometer, which is filled with mercury to serve as a thermal conductor between the surroundings and the reservoir. Besides protecting the thermometer from damage the tube is an essential part of the device because it eliminates the effect of hydrostatic pressure.

Readings obtained by reversing thermometers must be corrected for calibration errors and for the changes due to differences between the temperature at reversal and the temperature at which they are read. An equation developed by Schumacher is commonly used for this purpose:

$$\Delta T = \left[\frac{(T' - t)(T' + V_0)}{K}\right]\left[1 + \frac{(T' + t)(T' + V_0)}{K}\right] + I.$$

Here ΔT is the correction to be added algebraically to the uncorrected reading of the reversing thermometer, T'. The temperature at which the instrument is read is t, V_0 is the volume of the small bulb and of the capillary up to the 0°C graduation in terms of degree units, and K is a constant depending upon the relative thermal expansion of mercury and the type of glass used in the thermometer. For most reversing thermometers, K equals 6100. The term I is the calibration correction, which varies with the value of T'. Where there are large numbers of observations to be corrected, it is convenient to prepare graphs or tables for each thermometer from which the value of ΔT can be obtained for any values of T' and t, and where the calibration correction is included.

Reversing thermometers are generally used in pairs. The frames that hold them are brass tubes which have been cut away to make the scale visible and which are perforated around the reservoir. To hold the thermometers firmly without subjecting them to strain, the ends of the tubes are fitted with coil springs packed with sponge rubber.

UNPROTECTED REVERSING THERMOMETERS. Reversing thermometers, identical in design but mounted in open glass tubes, are employed to determine the depths of sampling, because thermometers subjected to pressure give a fictitious "temperature" reading that is dependent upon the temperature and the pressure. The unprotected reversing thermometers are so designed that the apparent temperature increase due to the hydrostatic pressure is about 0.01° per meter. An unprotected thermometer is always paired with a protected thermometer, by means of which the water temperature *in situ*, T_w, is determined. When T_w has been obtained by correcting the readings of the protected thermometer, the correction to be added to the reading of the unprotected thermometer can be obtained from the equation

$$\Delta T_u = \frac{(T_w - t_u)(T'_u + V_{0,u})}{K} + I_u.$$

Here T'_u and t_u are the readings of the unprotected reversing thermometer and its auxiliary thermometer, and I_u is the calibration correction. The difference between the corrected reading of the unprotected thermometer, T_u, and the corrected reading of the protected thermometer, T_w, represents the effect of the hydrostatic pressure at the depth of reversal. The depth of reversal is calculated from the expression

$$D \text{ (meters)} = \frac{T_u - T_w}{q\rho_m},$$

where q is the pressure constant for the individual thermometer. This constant is expressed in degrees increase in apparent temperature due to a pressure of 0.1 kg/cm². ρ_m is the mean density *in situ* of the overlying water. For work within any limited area it is usually adequate to establish a set of standard mean densities for use at different levels. Depths obtained by means of unprotected thermometers are of the greatest value when the wire rope holding the thermometers is not vertical in the water. When serial observations are made (p. 43), unprotected thermometers are usually placed on the lowest sampling bottle and, if possible, one on an intermediate bottle and one on a bottle near the top of the cast. The accuracy of depths obtained by unprotected thermometers depends upon the accuracy of the pressure constant and upon the accuracy of the readings of the two thermometers. The probable error is about ±5 m for depths less than 1000 m, and at greater depths is about 0.5 per cent of the wire depth.

THERMOGRAPHS. Many devices have been suggested for obtaining continuous observations at selected levels or as a function of depth. Thermographs are commonly used at shore stations and on vessels to obtain a continuous record at or near the sea surface. On board ship the thermometer bulb, usually containing mercury, is mounted on the ship's hull or in one of the intake pipes and connected to the recording mechanism by a fine capillary. The recording mechanism traces the temperature on a paper-covered, revolving drum. Records of temperatures obtained by a thermograph should be checked at frequent intervals against temperatures obtained in some other way.

Various types of electrical resistance thermometers have been designed to be lowered into the water and to give a continuous reading, but these devices have not proved satisfactory. Recently Spilhaus has developed an instrument called a *bathythermograph*, which can be used to obtain a record of temperature as a function of depth in the upper 150 m, where the most pronounced vertical changes are usually found. Essentially, the instrument is similar to a Jaumotte barograph. The temperature-sensitive part activates a bourdon-type element which moves a pen resting against a small smoked-glass slide which, in turn, is moved by a pressure-responsive element. As the instrument is lowered into the water and again raised, the pen traces temperature against pressure (hence, depth). This device has the great advantage that it can be operated at frequent intervals for obtaining a very detailed picture of the temperature distribution in the upper 150 m while the vessel is under way. Mosby has devised an instrument called a thermo-sounder for measuring temperature against depth which can be used for observations to great depths. The thermal element, mounted on an invar steel frame, is a 75-cm length of specially treated brass wire attached to a pen which makes a trace upon a circular slide that is slowly turned by means of a propeller as the instrument is lowered through the water.

Water-Sampling Devices

Water-sampling devices are of two general types; in one the closing is accomplished by means of plug valves, and in the other the closing is accomplished by plates seated in rubber. The *Nansen bottle* is an example of the first type and is the one most widely used in oceanographic research. The *Ekman bottle* is of the second type.

The Nansen bottle (fig. 5) is a reversing bottle fitted with two valves and holds about 1200 ml. The two plug valves, one on each end of the brass cylinder, are operated synchronously by means of a connecting rod that is fastened to the clamp securing the bottle to the wire rope. When the bottle is lowered, this clamp is at the lower end and the valves are in the open position so that the water can pass through the bottle. The bottle is held in this position by the release mechanism, which passes around the wire rope, but, when a messenger weight is sent down the rope and strikes the release, the bottle falls over and turns through 180 degrees, shutting the valves, which are then held closed by a locking device, and

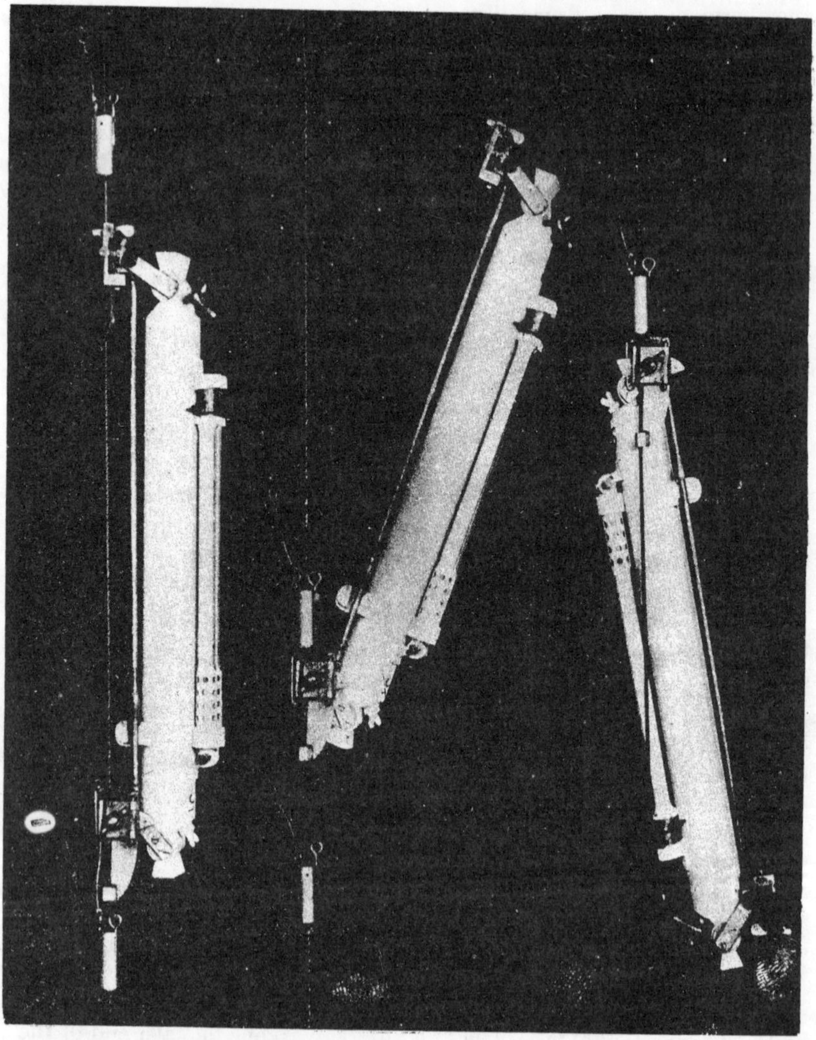

Fig. 5. For descriptive legend see opposite page.

reversing the attached thermometers. A number of bottles can be attached to the same wire rope. After reversing the first bottle, the messenger releases another messenger that was attached to the wire clamp before lowering. This second messenger closes the next lower bottle, releasing a third messenger, and so on.

The Ekman bottle, which can also be operated in series, consists of a cylindrical tube and top and bottom plates that are fitted with rubber gaskets. The moving parts are suspended in a frame that is attached to the wire rope, and, when the bottle is lowered, the water can pass freely through the cylinder. When struck by a messenger, the catch is released and the cylinder turns through 180 degrees, thereby pressing the end plates securely against the cylinder and enclosing the water sample. Reversing thermometers are mounted on the cylinder. In recent years the Nansen bottle has been generally used because of its more reliable functioning.

Messengers are essential for the operation of many types of oceanographic equipment. Although their size and shape will vary for different types of apparatus, they are essentially weights that are drilled out so that they will slide down the wire rope. In order to remove and attach them they are either hinged or slotted.

Treatment and Analysis of Serial Observations

Subsurface temperature and salinity observations obtained must be carefully examined for possible errors before they are arranged into convenient form for analyses, computations, or comparisons with other data. First of all, the depths of sampling have to be determined. This involves a considerable amount of practical experience, since depths obtained by unprotected thermometers may be in error because of improper functioning of the instruments, and also because unprotected thermometers are generally attached only to two or three water bottles of a string of bottles on a wire rope, the curvature of which is unknown.

When the depths have been found, vertical distribution curves are plotted by which temperature and salinity values that appear doubtful may be readily recognized. Temperatures may be in error because of faulty functioning of the thermometers or because the thermometers have been reversed prematurely, and salinities may be in error because the water bottles have closed prematurely or have leaked when being hauled up.

Besides plotting the vertical distribution curves, it is very helpful to plot the temperature-salinity curve (*T-S* curve) in which the corresponding values of temperature and salinity from a single station are entered in

---

Fig. 5. The Nansen reversing water bottle. *Left:* Before reversing; first messenger approaches releasing mechanism. *Center:* Bottle reversing; first messenger has released the second. *Right:* In reversed position.

a graph, with temperature and salinity as the coordinates, and are joined by a curve in order of increasing depth (p. 87). In any given area the shape of the *T-S* curves has a fairly definite form, and hence errors in observations may sometimes be detected from such a graph.

The interpolated values of temperature and salinity at standard depths are read off before the density, specific volume anomaly, and other calculations based upon these data are made (see chapter II). As a check on the correctness of the data, it is often advisable to plot the vertical distribution of density (σ_t) and the specific volume anomalies for each station as functions of depth.

Current Measurements

UNITS AND TERMS. In scientific literature the velocity of a current is given in centimeters per second (cm/sec) or occasionally in meters per second (m/sec), but in publications on navigation the velocity is stated in knots (nautical miles per hour) or in nautical miles per 24 hours. The direction is always given as the direction toward which the current flows, because a navigator is interested in knowing the direction in which his vessel is carried by the current. The direction is indicated by compass points (for example, NNW, SE), by degrees reckoned from north or south toward east or west (for example, N 60° W, S 30° E), or in degrees from 0° to 360°, counting current toward north as 0° (or 360°) and current toward south as 180°.

DRIFT METHODS. Information as to the general direction of surface currents is obtained from the drift of floating objects such as logs, wreckage from vessels, and fishermen's implements. In most instances, conclusions as to currents from the finding of accidental drifting bodies are incomplete because the locality and the time at which the drift started are not known, nor is it known how long the object might have been lying on the beach before discovery. Neither is it known to what extent such drifting bodies have "sailed" through the water, being driven forward by winds.

More than a century ago, in order to overcome such uncertainties, *drift bottles* were introduced. These are weighted down with sand so that they will be nearly immersed, offering only a very small surface for the wind to act on, and they are carefully sealed. They contain cards giving the number of the bottle, which establishes the locality and time of release, and requesting the finder to fill in information as to place and time of finding.

The *interpretation* of results of drift-bottle experiments presents difficulties. In general, a bottle does not follow a straight course from the place of release to the place of finding, and conclusions as to the probable drift must be guided by knowledge of the temperature and salinity distribution in the surface layers. Fairly accurate estimates of the average speed of the drift can be made if the bottle is picked up from the water.

Drift bottles have been used successfully for obtaining information as to surface currents over relatively large ocean areas, such as the equatorial part of the Atlantic Ocean and the seas around Japan. They have supplied numerous details in more enclosed seas like the English Channel and the North Sea, but have proved less successful off an open coast.

The drift method can also be used for obtaining information as to currents in a shorter time interval. By far the greatest number of observations of surface currents have been derived from ships' records, and these data are obtained by the drift method. On board ship the position of a vessel, weather permitting, is determined by astronomic observations of the sun or stars. From this position, indicated by *A* in fig. 6, the course along which to continue is decided upon, but, when a new "fix" is obtained (*B′* in fig. 6), it is in general found that the vessel is not at the location *B*, where it should be according to the course that has been steered, taking the wind drift of the vessel into account, and the distance covered according to the log (the position by "dead reckoning"). The displacement is considered due to currents in the time interval between "fixes." The method for computing these currents is shown in fig. 6. This method gives, in general, the average current in twenty-four hours or multiples of twenty-four hours.

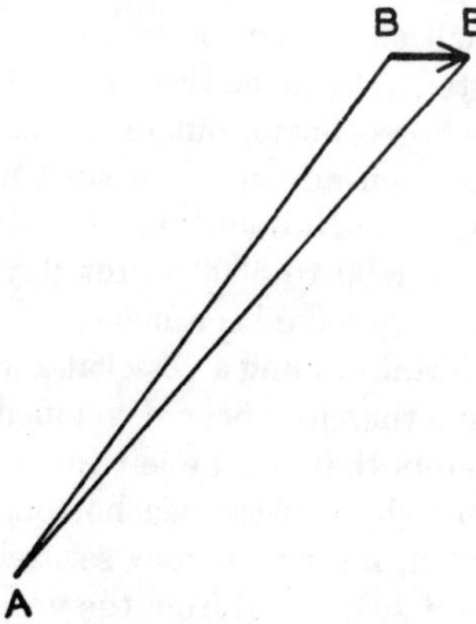

Fig. 6. Determination of surface currents by difference between positions by fixes and dead reckoning.

From an anchored vessel, say a lightship, the surface current can be determined either by a chip log or by drift buoys, below which, in general, is a "current cross" acting as a sea anchor. This type of drift buoy was used on the *Challenger*. The latter methods give nearly instantaneous values of the surface currents at the place of observation.

Near land the methods can be elaborated in such a manner that the drift of a body can be determined in detail over long distances and long periods. A drifting buoy can be followed by a vessel whose positions can be accurately established by bearings on known landmarks, or the buoy can be provided with a mast, and the direction to the buoy can be observed and its distance from a fixed locality can be measured by a range finder. Both methods have been used successfully. The latter can also be employed in the open ocean by anchoring one buoy, setting another buoy adrift, and determining the bearing of and the distance to the drifting buoy from a ship that remains as close as possible to the anchored buoy.

Flow Methods. From an anchored vessel or float the currents can be measured by stationary instruments past which the current flows, turning a propeller of some type or exerting a pressure which can be determined by different methods. The advantage of these methods is that observations need not be limited to the currents of the surface layers but can be extended to any depth. The obvious difficulty is to retain the instrument in a fixed locality so that the absolute flow of the water may be measured, and not merely the flow relative to a moving instrument. In shallow water a vessel can be anchored so that the motion of the vessel is small enough to be insignificant or of such nature that it can be eliminated. In deep water, current measurements were first made from anchored boats, but in recent years the technique of deep-sea anchoring has been advanced to such an extent that vessels like the *Meteor*, the *Armauer Hansen*, and the *Atlantis* have remained anchored in depths from 4000 to 5500 m for days and weeks. In other instances, relative currents have been measured from slowly drifting vessels.

Maintaining a vessel at anchor for a long time is expensive, and devices have therefore been developed for anchoring automatic recording current meters that can be left for weeks at a time. Measurements of currents very close to the sea bottom cannot be made safely from an anchored vessel, no matter how securely it is kept in position, because an instrument suspended from the vessel cannot be retained at a constant distance from the bottom owing to the motion due to swells and tides. The difficulty was first overcome by Nansen, who lowered a tripod to the sea bottom and suspended a current meter from the top of the tripod. This method has more recently been used by Stetson, by Revelle and Fleming, and by Revelle and Shepard. The latter suspended three current meters from the top of the tripod and were thus able to obtain simultaneous measurements of currents at three levels within less than 2 m of the bottom.

Numerous types of current meters have been designed, each having special advantages and disadvantages. Owing to its simplicity and reliability, the *Ekman current meter* is widely used (fig. 7). The essential parts of the instrument are the propeller, the revolutions of which are recorded on a set of dials, the compass box with the device for recording the orientation of the meter, and the vane that orients the instrument so that the propeller faces the current. The free swing of the instrument is ensured by mounting it in ball bearings on a vertical axis. The wire for lowering is fastened to the upper end of this axis, and a suitable weight is attached below the axis. The instrument is balanced in water so that the axis is vertical. The carefully balanced propeller, with four to eight thin, light blades, runs inside a strong protective ring, but can easily be removed for inspection or transportation. The axis of the propeller runs with tantalum points on agate bearings. Inside the protective ring is a lever that can be operated by messengers. With the lever in its lowest position the propeller is arrested, and in this state the instrument is lowered. When the desired depth is reached, a messenger weight is dropped, which pushes the lever up to its middle position, releasing the propeller, the turns of which are recorded on a set of dials. After a number of minutes a second messenger weight is dropped, which pushes the lever up to the highest position, stopping the propeller. In later types the propeller is also shielded in front when the instrument is

Fig. 7.
The Ekman current meter. Messenger, dials, and compass box are seen. Propeller is hidden by protective ring.

lowered in order to prevent fouling of the propeller by such organisms as medusae. This front shield is opened by the first messenger.

The direction of the current is recorded by an ingenious device which is simple and reliable. A tube extends from above the dial box to a disk on the axis of the cogwheel, which turns once when the propeller makes one hundred revolutions. This tube is filled with phosphor-bronze balls about 2 mm in diameter. In the disk are three or more indentations corresponding to the size of the balls. When one of these indentations passes below the tube containing the balls, a ball drops into it and is carried around with the disk until it drops into a second tube which extends downward and ends above the center of the compass box. In the compass box, which can be easily removed from the bar to which it is fastened, a system of magnets swings freely on a pin that runs on agate. The frame to which the magnets are fastened carries a bar that is shaped like a wide, inverted V. The upper side of one arm of the bar is trough-shaped, so that a ball dropping through a hole in the center of the lid of the box runs down this trough and falls to the bottom of the box, which is divided into thirty-six compartments, each corresponding to an angle of 10 degrees and marked N, N 10° E, N 20° E, and so on. The compass box is rigidly connected to the vane of the meter, but the magnets of the compass adjust themselves in the magnetic meridian. The compartment into which the ball drops therefore indicates the direction of the vane—that is, the direction of the current at the moment the ball fell. In general, several balls fall during one observation, since one ball drops for each thirty-three revolutions or less of the propeller. The average direction of the current is obtained by computing the weighted mean according to the distribution of the balls. If the direction has varied widely during the period of observation, the average direction will be uncertain or even indeterminate, in which case the average velocity as computed from the revolution of the propeller has no meaning.

CHAPTER IV

The Heat Budget of the Oceans

To the heat budget of the earth as a whole, only processes of radiation are of importance, but the heat budget of the atmosphere is more complicated, being controlled by processes of radiation, condensation of water vapor, and exchange of sensible heat with the ocean and land surfaces. The heat budget of the oceans is primarily controlled by processes of radiation, exchange of sensible heat with the atmosphere, and evaporation from the surface or condensation of water vapor on the surface, but such processes as convection of heat through the ocean bottom from the interior of the earth, transformation of kinetic energy into heat, and heating due to chemical processes also have to be examined.

The processes involved can be listed as follows:

Processes of heating of the ocean water	*Processes of cooling of the ocean water*
1. Absorption of radiation from the sun and the sky, Q_s	1. Back radiation from the sea surface, Q_b
2. Convection of heat through the ocean bottom from the interior of the earth	2. Convection of sensible heat to the atmosphere, Q_h
3. Transformation of kinetic energy to heat	3. Evaporation, Q_e
4. Heating due to chemical processes	
5. Convection of sensible heat from the atmosphere	
6. Condensation of water vapor.	

Over all oceans between 70° N and 70° S the average incoming radiation from the sun and the sky which penetrates the sea surface is about 0.22 g cal/cm²/min, and amounts that are small in comparison to this can be neglected.

It has been estimated that the flow of heat through the bottom of the sea amounts to between 50 and 80 g cal/cm²/year—that is, 0.095×10^{-3} to 0.152×10^{-3} g cal/cm²/min. This amount represents less than one thousandth part of the radiation received at the surface and can be neglected when dealing with the heat budget of the oceans. In a few basins where the deep water is nearly stagnant and where conduction of heat from above or from the sides is negligible, the amount of heat conducted through the bottom may conceivably play a part in determining the distribution of temperature, but so far no such case is known with certainty.

3. The kinetic energy transmitted to the sea by the stress of the wind on the surface and part of the tidal energy are dissipated by friction and transformed into heat. The energy transmitted by the wind can be estimated at about one ten-thousandth part of the radiation received at the surface and can be neglected. In shallow coastal waters with strong tidal currents the dissipation of tidal energy may be so great, however, that it may become of some local importance. Thus, in the Irish Channel, according to Taylor, the dissipation amounts to about 0.002 g cal/cm^2/min, or 1050 g cal/cm^2/year. The average depth can be taken as about 50 m, or 5000 cm, and, if the same water remained in the Irish Channel a full year, the increase in temperature would be about 0.2° C, on an average. Such an effect, however, has not been established, and, as it can be expected in shallow coastal waters only, it is of no significance to the general heat budget of the oceans.

4. It is estimated that in ocean regions of abundant plant life, up to 0.8 per cent of the incoming radiation may be utilized by the plants for photosynthesis, but over all ocean areas the average amount is probably less than one tenth of the maximum values and can be neglected as unimportant.

The only processes to be considered, therefore, are the radiation processes and the exchange of heat and water vapor with the atmosphere, so that for the oceans as a whole the average annual heat budget can be written in the form

$$Q_s - Q_b - Q_h - Q_e = 0. \qquad \text{(IV, 1)}$$

If specific regions and time intervals are considered, it must be taken into account that heat may be brought into or out of a region by ocean currents or by processes of mixing, and that during short time intervals a certain amount of heat may be used for changing the temperature of the water. The complete equation for the heat balance of any part of the ocean in a given time interval is, therefore,

$$Q_s - Q_b - Q_h - Q_e - Q_v - Q_\vartheta = 0, \qquad \text{(IV, 2)}$$

where Q_v represents the net amount that is brought into or out of the region by currents or processes of mixing, and where Q_ϑ represents the amount of heat used locally for changing the temperature of the sea water.

Radiation

INCOMING RADIATION; EFFECT OF CLOUDS; REFLECTION. Part of the short-wave radiation that reaches the sea surface comes directly from the sun and part of it comes from the sky as reflected or scattered radiation. The amount of radiation energy that is absorbed per unit volume in the sea depends upon the amount of energy which reaches the sea surface, the reflection from the sea surface, and the extinction coefficients for total energy. The incoming radiation depends mainly upon the altitude of the sun, the absorption in the atmosphere, and the cloudiness. With a clear sky and a high sun, about 85 per cent of the radiation comes directly from the sun and about 15 per cent from the sky, but with a low sun the proportion from the sky is greater, reaching about 40 per cent of the total with the sun 10 degrees above the horizon.

The incoming energy from the sun is cut down in passing through the atmosphere, partly owing to absorption by water vapor and carbon dioxide in the air, but mainly by scattering against the air molecules or very fine dust. The total effect of absorption and scattering in the atmosphere depends upon the thickness of the air mass through which the sun's rays pass, as expressed by the equation

$$I = Se^{-Ba_m m}. \qquad \text{(IV, 3)}$$

Here I represents the energy in g cal/cm²/min reaching a surface that is normal to the sun's rays; m represents the relative thickness of the air mass and is equal to 1 at a pressure of 760 mm when the sun stands in zenith, equal to 2 when the sun is 30° above the horizon (sin 30° = ½), and so on; S is the solar constant and is equal to 1.94 g cal/cm²/min; B is the "turbidity factor" of the air; and $a_m = 0.128 - 0.054 \log m$.

The sun's radiation on a horizontal surface is obtained by multiplication with sin h, where h is the sun's altitude. To this amount must be added the diffuse sky radiation in order to obtain the total radiation on a horizontal surface. Instruments are in use for recording the total radiation and for recording separately the radiation from the sun and the sky.

When the sun is obscured by clouds, the radiation comes from the sky and the clouds, and on an average can be represented by the formula $Q = Q_0(1 - 0.071C)$, where the cloudiness C is given on the scale 0 to 10, and where Q_0 represents the total incoming radiation with a clear sky. This formula is applicable, however, only to average conditions. If the sun shines through scattered clouds, the radiation may be greater than with a clear sky, owing to the reflection from the clouds, and on a completely overcast, dark and rainy day the incoming radiation may be cut down to less than 10 per cent of that on a clear day. Table 6 contains the average monthly amounts of incoming radiation, expressed in gram

TABLE 6

AVERAGE AMOUNTS OF RADIATION FROM SUN AND SKY, EXPRESSED IN GRAM CALORIES PER SQUARE CENTIMETER PER MINUTE, WHICH EVERY MONTH REACHES THE SEA SURFACE IN THE STATED LOCALITIES
(After Kimball)

Locality		Month											
Latitude	Longitude	Jan.	Feb.	Mar.	Apr.	May	June	July	Aug.	Sept.	Oct.	Nov.	Dec.
60°N	7°E- 56°W	.002	.053	.125	.207	.272	.292	.267	.212	.147	.074	.006	0
60 N	135 -170 W	.005	.078	.155	.208	.269	.260	.242	.185	.127	.077	.015	0
52 N	10 W	.048	.089	.148	.219	.258	.267	.251	.211	.160	.104	.062	.041
52 N	129 W	.053	.091	.135	.185	.246	.250	.230	.214	.158	.097	.058	.039
42 N	66 - 70 W	.094	.138	.212	.272	.306	.329	.302	.267	.230	.174	.115	.086
42 N	124 W	.100	.151	.210	.286	.331	.360	.320	.274	.231	.174	.113	.092
30 N	65 - 77 W	.146	.165	.238	.285	.317	.310	.301	.282	.239	.188	.169	.142
30 N	128 -130 E	.141	.153	.199	.241	.258	.238	.256	.260	.219	.178	.153	.135
10 N	61 - 69 W	.254	.276	.299	.305	.272	.276	.285	.292	.287	.269	.248	.239
10 N	116 E- 80 W	.226	.257	.292	.278	.255	.239	.240	.242	.247	.237	.224	.219
0	7 - 12 E	.239	.248	.244	.230	.210	.196	.188	.194	.220	.240	.239	.235
0	48 W & 170 E	.261	.265	.282	.297	.309	.300	.300	.340	.366	.362	.339	.278
10 S	14 E; 36- 38 W	.329	.328	.301	.254	.219	.206	.232	.278	.312	.324	.317	.320
10 S	72 -171 E	.290	.308	.315	.289	.266	.253	.269	.306	.332	.313	.301	.303
30 S	17 and 116 E	.452	.406	.340	.254	.186	.148	.166	.214	.274	.362	.401	.430
30 S	110 W	.380	.330	.260	.209	.162	.130	.145	.176	.237	.321	.340	.390
42 S	73 W; 147 E	.343	.297	.223	.154	.104	.085	.092	.135	.187	.264	.310	.348
52 S	58 W	.289	.237	.167	.112	.062	.039	.049	.097	.150	.222	.273	.302
60 S	45 W	.213	.171	.105	.056	.011	0	.003	.054	.111	.156	.204	.221

calories per square centimeter per minute, which reach a horizontal surface in the indicated localities (computed from Kimball). The differences between the parts of the oceans in the same latitudes are mainly due to differences in cloudiness.

Few direct measurements of radiation are available from the oceans, and when dealing with the incoming radiation it is necessary to consider average values that can be computed from empirical formulas. Mosby has established such a formula by means of which monthly or annual mean values of the incoming radiation on a horizontal surface can be computed if the corresponding average altitude of the sun and the average cloudiness are known.

$$Q = k\bar{h}(1 - 0.071C) \text{ g cal/cm}^2\text{/min.} \qquad \text{(IV, 4)}$$

Here $\bar{h}$ is the average altitude of the sun. The factor k depends upon the transparency of the atmosphere and appears to vary somewhat with latitude, being 0.023 at the Equator, 0.024 in lat. 40°, and 0.027 in lat. 70°. The values computed by means of this formula agree within a few per cent with those derived by Kimball in an entirely different manner (table 6).

Part of the incoming radiation is lost by reflection from the sea surface, the loss depending upon the altitude of the sun. When computing the loss, the direct radiation from the sun and the scattered radiation from the sky must be considered separately. With the sun 90°, 60°, 30°, and 10° above the horizon, the reflected amounts of the direct solar radiation would be, according to Schmidt, 2.0 per cent, 2.1 per cent, 6.0 per cent, and 34.8 per cent, respectively. For diffuse radiation from the sky and from clouds, Schmidt computes a reflection of 17 per cent. Measurements made by Powell and Clarke give values on clear days in agreement with the above, but on overcast days, when all radiation reaching the sea surface was diffuse, the observed reflection was about 8 per cent. If the fractions of the total radiation from the sun and the sky on a clear day are called p and q, respectively, and if the corresponding percentages reflected are called m and n, the percentage of the total incoming radiation which is reflected on a clear day is $r = mp + nq$. On an overcast day, when all incoming radiation is diffused, $r = 8$ per cent. Table 7 shows approximate values of r at different altitudes of the sun on a clear day.

TABLE 7

PERCENTAGE OF TOTAL INCOMING RADIATION FROM SUN AND SKY WHICH ON A CLEAR DAY IS REFLECTED FROM A HORIZONTAL WATER SURFACE AT DIFFERENT ALTITUDES OF THE SUN

Altitude of the sun	5°	10°	20°	30°	40°	50°	60°	70°	80°	90°
Percentage reflected	40	25	12	6	4	3	3	3	3	3

The values in the table are applicable only if the sea surface is smooth. In the presence of waves the reflection loss at a low sun is somewhat increased and will be of particular importance in high latitudes. The amount of radiation which under stated conditions penetrates the sea surface is obtained by subtracting the reflection loss from the total incoming radiation.

ABSORPTION OF RADIATION ENERGY IN THE SEA. The radiation that penetrates the surface is absorbed in the sea water. The amounts absorbed within given layers of water can be derived by measuring with a thermopile the energies that reach different depths or by computing these

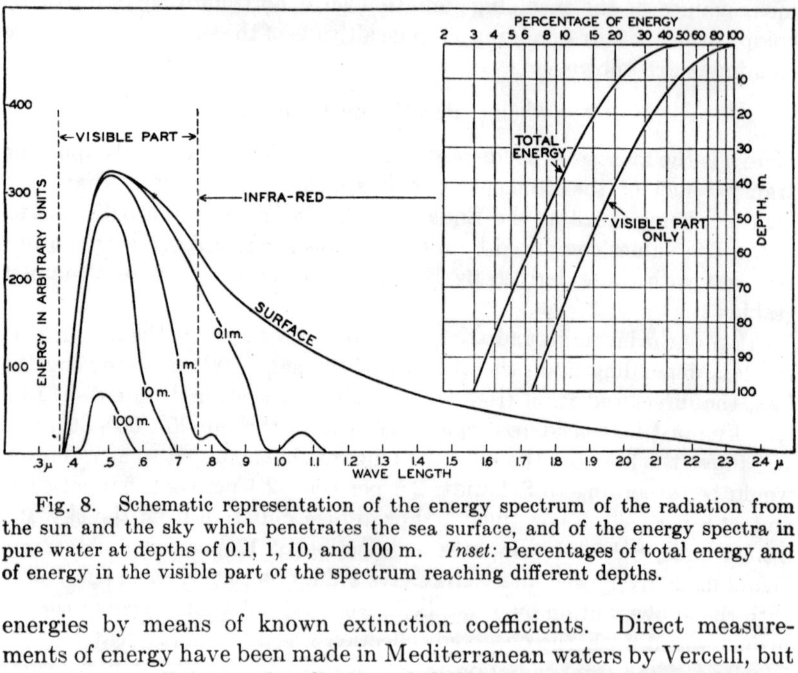

Fig. 8. Schematic representation of the energy spectrum of the radiation from the sun and the sky which penetrates the sea surface, and of the energy spectra in pure water at depths of 0.1, 1, 10, and 100 m. *Inset:* Percentages of total energy and of energy in the visible part of the spectrum reaching different depths.

energies by means of known extinction coefficients. Direct measurements of energy have been made in Mediterranean waters by Vercelli, but extinction coefficients of radiation of different wave lengths have been determined in many areas (p. 30). For computation of the energy which reaches a given depth it is necessary to know the intensity of the radiation at different wave lengths—that is, the energy spectrum. The reduction in intensity has to be calculated for each wave length, and the total energy reaching a given depth has to be determined from the energy spectrum by means of integration. The definition of the extinction coefficient for total energy corresponds to the definition of extinction coefficients at given wave lengths (p. 27).

The spectrum of the energy that penetrates the sea surface is represented approximately by the upper curve in fig. 8, which also shows the energy spectra at different depths in pure water. The total energy at any given depth is proportional to the area enclosed between the base line and the curves showing the energy spectrum. In the inserted diagram the total energy, expressed as percentage of the energy penetrating the surface, as well as the corresponding percentages of the energy in the visible part of the spectrum, are plotted against depth. The figure shows that pure water is transparent for visible radiation only.

For sea water the percentage of the total energy reaching various depths has been computed for the clearest oceanic water, for average oceanic water, for average coastal water, and for turbid coastal water, using the extinction coefficients shown in table 4. The results are presented in table 8. In even the clearest offshore water 62.3 per cent of the incoming energy is absorbed in the first meter. The absorption is often increased in the upper one meter because of the presence of foam

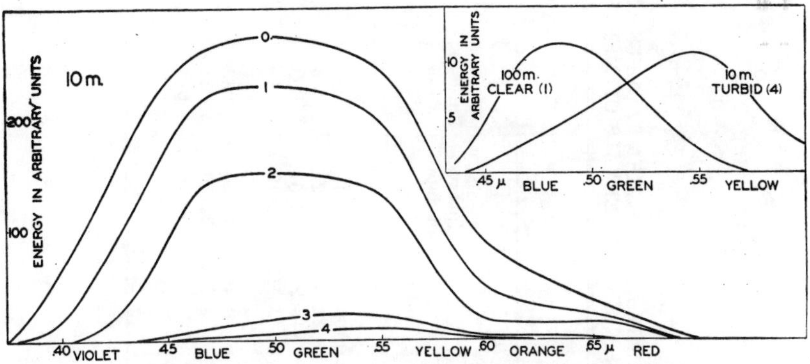

Fig. 9. Energy spectra at a depth of 10 m in different types of water. Curves marked 0, 1, 2, 3, and 4 represent energy spectra in pure water, clear oceanic, average oceanic, average coastal, and turbid coastal sea water, respectively. *Inset:* Energy spectra at a depth of 100 m in clear oceanic water and at 10 m in turbid coastal water.

and air bubbles. This increased absorption, when dealing with the penetration of light, is referred to as "surface loss." If this process is disregarded, the values clearly demonstrate that the greater amount of energy is absorbed very near the sea surface and that the amount which penetrates to any appreciable depth is considerable only when the water is exceptionally clear. At 10 m, 83.9 per cent has been absorbed in the clearest water and 99.55 per cent in the turbid coastal water.

The absorption of energy is illustrated in fig. 9, which shows the energy spectra in different types of water at a depth of 10 m. At this depth the maximum energy in the clearest water is found in the blue-green portion of the spectrum, whereas in the turbid coastal water the maximum has been displaced toward the greenish-yellow part. This displacement is further illustrated by the inserted curve in the upper right-hand corner of the figure, which shows the energy spectra at 100 m in the clearest water and at 10 m in the most turbid water.

TABLE 8

PERCENTAGE AMOUNTS OF TOTAL INCIDENT ENERGY AT DIFFERENT LEVELS AND EXTINCTION COEFFICIENTS OF ENERGY IN DIFFERENT INTERVALS OF DEPTH

Depth	Percentage amounts of incident energy					Interval of depth	Extinction coefficients per meter				
	Pure sea water	Oceanic water		Coastal water			Pure sea water	Oceanic water		Coastal water	
		Clearest	Average	Average	Turbid			Clearest	Average	Average	Turbid
0	100	100	100	100	100	0–1	.944	.975	1.080	1.318	1.385
1	38.9	37.7	35.2	26.7	22.8	1–2	.143	.176	.230	.450	.547
2	33.7	31.6	28.0	17.0	13.2	2–5	.062	.095	.159	.351	.452
5	28.0	23.7	17.3	5.95	3.41	5–10	.048	.076	.120	.318	.405
10	22.0	16.1	9.50	1.21	0.449	10–20	.033	.054	.094	.292	.368
20	15.8	9.35	3.72	0.064	0.012	20–50	.024	.042	.083		
50	7.64	2.69	0.311			50–100	.018	.036			
100	3.04	0.452	0.0057								

Extinction coefficients of *total* energy have been computed and are entered in table 8. These extinction coefficients are very high in the upper one meter but decrease rapidly, at greater depth approaching the minimum extinction coefficients characteristic of the types of water dealt with. The smallest values given in the table can be considered valid at greater depths as well.

In fig. 10 the curves marked 0, 1, 2, 3, and 4 represent the percentage amounts of energy that reach different levels between the surface and 10 m, according to the data in table 8. The three curves marked Capri,

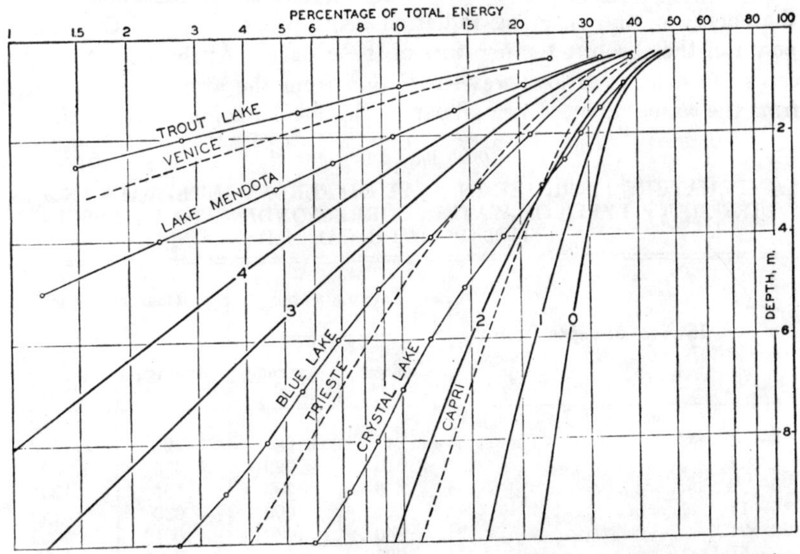

Fig. 10. Percentages of total energy reaching different depths in pure water, clear oceanic, average oceanic, average coastal, and turbid coastal sea water (curves 0, 1, 2, 3, and 4) computed from extinction coefficients and corresponding to directly observed values in four lakes and at three localities in the Mediterranean.

Trieste, and Venice represent results of measurements in the Mediterranean according to Vercelli, and four other curves represent observed values in lakes according to Birge and Juday. The agreement of the character of the curves indicates that one can arrive at reliable values as to the absorption of energy in the sea by means of calculations based on extinction coefficients.

An idea of the heating due to absorption of radiation can be obtained by computing the increase of temperature at different depths which results from a penetration of 1000 g cal/cm² through the surface. The results are shown in table 9, which serves to emphasize the fact that the greater part of the energy is absorbed near the surface, particularly in turbid water. If no other processes took place, the temperature between the surface and 1 m would increase in the clearest water by 6.24°, and in the most turbid water, by 7.72°. Between 20 and 21 m the corresponding values would be 0.04° and 0.0003°.

The temperature changes recorded in table 9 show no similarity to those actually occurring in the open oceans, where processes of mixing entirely mask the direct effect of absorption, but in some small, land-locked bodies of water the temperature changes at subsurface depths may be governed mainly by absorption of short-wave radiation.

Effective Back Radiation from the Sea Surface. The sea surface emits long-wave heat radiation, radiating nearly like a black body, the energy of the outgoing radiation being proportional to the fourth power of the absolute temperature of the surface. At the same time the sea surface receives long-wave radiation from the atmosphere, mainly from the water vapor. A small part of this incoming long-wave radiation

Table 9

TEMPERATURE INCREASE IN °C AT DIFFERENT INTERVALS AND IN DIFFERENT TYPES OF WATER, CORRESPONDING TO AN ABSORPTION OF 1000 G CAL/CM²

Interval of depth	Oceanic water		Coastal water	
	Clearest	Average	Average	Turbid
0– 1	6.24	6.48	7.32	7.72
1– 2	0.610	0.720	0.970	0.960
5– 6	.236	.282	.164	.120
10– 11	.104	.096	.030	.0140
20– 21	.040	.030	.0016	.0003
50– 51	.0096	.0024	$.0_5 34$	$.0_7 15$
100–101	.0016	$.0_4 11$		

is reflected from the sea surface, but the greater portion is absorbed in a small fraction of a centimeter of water, because the absorption coefficients are enormous at long wave lengths. The *effective back radiation* from the sea surface is represented by the difference between the "temperature radiation" of the surface and the long-wave radiation from the atmosphere, and this effective radiation depends mainly upon the temperature of the sea surface and the water-vapor content of the atmosphere. According to Ångstrom the latter can be put proportional to the local vapor pressure, which can be computed from the relative humidity if the air temperature is known. Over the oceans the air temperature deviates so little from the sea-surface temperature that the vapor pressure can be obtained with sufficient accuracy from the sea-surface temperature and the relative humidity of the air at a short distance above the surface.

Ångstrom has published a table that summarizes the results of observations of effective radiation against a clear sky from a black body of different temperatures and at different vapor pressures. Fig. 11 has been prepared by means of this table, taking into account the small differences between the radiation of a black body and that of a water surface. The figure shows the effective radiation as a function of sea-surface temperature and of relative humidities between 100 per cent and 70 per cent, but the values that can be read off from the graph may be 10 per cent in error, owing to the scanty information upon which the curves are based. It brings out the fact, however, that, owing to the increased radiation from the atmosphere at higher temperatures (higher vapor pressure), the effective back radiation *decreases* slowly with *increas-*

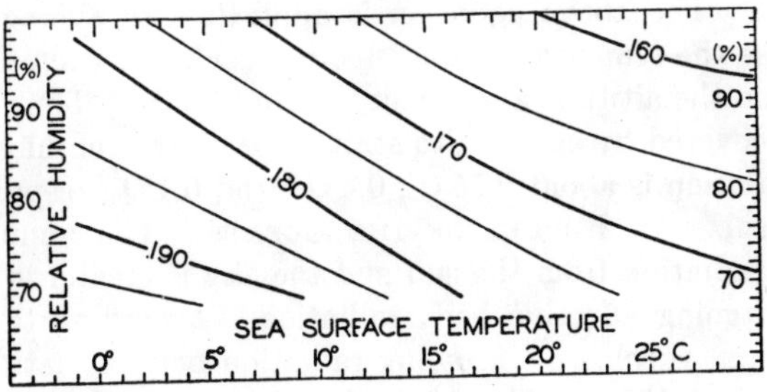

Fig. 11. The effective back radiation in gram calories per square centimeter per minute from the sea surface to a clear sky, represented as a function of sea surface temperature and relative humidity of the air at a height of a few meters.

ing temperature. At a temperature of 0°C and a relative humidity of 80 per cent, the effective back radiation decreases with increasing humidity, owing to the increasing back radiation from the atmosphere. Thus, at a surface temperature of 15° the effective radiation is about 0.180 g cal/cm²/min at a relative humidity of 70 per cent, and about 0.163 g cal/cm²/min at a relative humidity of 100 per cent.

The values of the effective back radiation at higher temperatures as obtained by extrapolation of Ångstrom's data (fig. 11) are greater than those computed from Brunt's empirical formula

$$Q_b = Q'(1 - 0.44 - 0.08\sqrt{e}),$$

where Q' is the radiation of a black body having the temperature of the sea surface and e is the vapor pressure of the air in millibars. However, in this formula the numerical values of the coefficients are uncertain and are applicable only within a range of e between 4 and 18 millibars.

The diurnal and annual variations of the sea-surface temperatures and of the relative humidity of the air over the oceans are small, and the effective back radiation at a clear sky is therefore nearly independent of the time of the day and of the season of the year, in contrast to the incoming short-wave radiation from the sun and the sky, which is subjected to very large diurnal and seasonal variations. In the presence of clouds the effective back radiation is cut down, because the radiation from the atmosphere is increased. The empirical relation can be written

$$Q = Q_0(1 - 0.083C), \qquad \text{(IV, 5)}$$

where Q_0 is the back radiation at a clear sky and where C is the cloudiness on the scale 1 to 10. A diurnal or annual variation in the cloudiness will lead to a corresponding variation in the effective back radiation. On an average, the diurnal variation of cloudiness over the oceans is very small and can be neglected, but the annual variation is in some regions considerable. The above equation is applicable to average conditions only, because the reduction of the effective back radiation due to clouds depends upon the altitude and density of the clouds. Thus, if the sky is completely covered by cirrus, alto stratus, or strato cumulus clouds, the effective radiation is about 0.75 Q_0, 0.4 Q_0, and 0.1 Q_0, respectively.

THE RADIATION BUDGET OF THE OCEANS. The annual incoming short-wave radiation from the sun and the sky is greater in all latitudes than the outgoing effective back radiation. According to Mosby the average annual surplus of incoming radiation between latitudes 0° and 10° N is about 0.170 g cal/cm²/min, and between 60° and 70° N, about 0.040 g cal/cm²/min. The surplus of radiation must be given off to the atmosphere, and the exchange of heat and water vapor with the atmosphere is therefore equally important with the processes of radiation in regulating the ocean temperature and salinity.

The characteristics of the oceans in respect to radiation are very favorable to man. The water surface reflects only a small fraction of the incoming radiation and the greater part of the radiation energy is absorbed in the water, distributed by processes of mixing over a layer of considerable thickness, and given off to the atmosphere during periods when the air is colder than the sea surface. The oceans therefore exercise a thermostatic control on climate. Conditions are completely changed, however, if the temperature of the sea surface decreases to the freezing point, so that further loss of heat from the sea leads to formation of ice, because when passing this critical temperature the characteristics are altered in a very unfavorable direction. Sea ice, which soon attains a gray-white appearance owing to enclosed air bubbles, reflects 50 per cent or more of the incoming radiation, and if covered by rime or snow the reflection loss increases to 65 per cent, or even to 80 per cent if covered by fresh, dry snow. The snow surface, on the other hand, radiates nearly like a black body, and consequently the heat budget related to processes of radiation, instead of rendering a surplus, as it does over the open ocean, shows a deficit until the temperature of the ice surface has been lowered so much that the decreased loss by effective back radiation balances the small fraction of the incoming radiation that is absorbed. The immediate result of freezing is therefore a general lowering of the surface temperature of the ice and a rapid increase of the thickness of the ice. The air that comes in contact with the ice is cooled, and, as this cold air spreads, more ice is formed. Thus, a small lowering of the temperature of the water in high latitudes followed by freezing may lead to a rapid drop of the air temperature and a rapid increase of the ice-covered area. On the other hand, a small increase of the temperature of air flowing in over an ice-covered sea may lead to melting of the ice at the outskirts and, once started, the melting may progress rapidly. In agreement with this reasoning, it has been found that the extent of ice-covered areas in the Barents Sea is a sensitive indicator of small changes in the atmospheric circulation and of small changes in the amount of warm water carried into the region by currents. It has also been computed that if the average air temperatures in middle and higher latitudes were raised a few degrees, the Polar Sea would soon become an ice-free ocean.

Exchange of Heat between the Atmosphere and the Sea

The amount of heat that in unit time is carried away from the sea surface through a unit area is equal to $-c_p\mu_e\left(\frac{d\vartheta}{dz}+\gamma\right)$, where c_p is the specific heat of the air, μ_e is the eddy conductivity, $-d\vartheta/dz$ is the temperature gradient of the air (the lapse rate), which is positive when the temperature decreases with height, and γ is the adiabatic lapse rate. Very near the sea surface, γ can be neglected as small compared to $d\vartheta/dz$. The term $c_p\mu_e$ enters instead of the coefficient of heat conductivity of the air as determined in the laboratory, because the air is nearly always in turbulent motion and because in the air $\mu_s = \mu_e$ (p. 19). The state of turbulence varies, however, with the distance from the sea surface, because at the surface itself the eddy motion must be greatly reduced. As a consequence, under steady conditions, when the same amount of heat passes upward through every cross section of a vertical column, the temperature changes rapidly with height near the sea surface and more slowly at a greater distance. The product $-c_p\mu_e d\vartheta/dz$ remains constant and, since $c_p\mu_e$ increases rapidly with height, $-d\vartheta/dz$ must decrease.

Detailed and accurate temperature measurements in the lowest meters of the air over the ocean have not yet been made, because the hull and masts of a vessel disturb the normal distribution of temperature to such an extent that values observed at different levels on board a vessel are not representative of the undisturbed conditions. The few measurements that have been attempted indicate, however, that the general distribution as outlined above is encountered.

The sea surface must be warmer than the air at a small distance above the surface if heat shall be conducted from the sea to the air. When such conditions prevail, the air is heated from below, the stratification of the air becomes unstable, and the turbulence of the air becomes intense. If the sea surface is very much warmer than the air, as may be the case when cold continental air flows out over the sea in winter, the heating from below may be so intense that rapid convection currents develop, leading to such violent atmospheric disturbances as thunderstorms. The point which is emphasized is that an appreciable conduction of heat from the sea to the atmosphere takes place when the sea surface is warmer than the air. One might assume that, vice versa, an appreciable amount of heat would be conducted to the sea surface when warmer air flows over a cold sea, but this is not the case, because under such conditions the air is cooled from below, the stratification of the air becomes stable, and the turbulence, and consequently the eddy conductivity of the air, is greatly reduced.

It has been found (p. 75) that on an average the sea surface is slightly warmer than the overlying air and therefore loses heat by conduction. So far, no detailed studies have been made, but Ångstrom has estimated that only about 10 per cent of the total heat surplus is given off to the atmosphere by conduction and that 90 per cent is used for evaporation. Other estimates indicate that these figures are approximately correct (p. 64). Thus, evaporation is of much greater importance to the heat balance of the oceans than is the transfer of sensible heat.

Evaporation from the Sea

The Process of Evaporation. The vapor tension at a flat surface of pure water depends on the temperature of the water. The salinity decreases the tension slightly, the empirical relation between vapor tension and salinity being (p. 15)

$$e_w = e_d(1 - 0.0053\, S),$$

where e_w is the vapor tension over sea water, e_d is the vapor tension over distilled water of the same temperature, and S is the salinity in parts per thousand. In the open ocean $e_w = 0.98e_d$, approximately.

In discussing the process of evaporation, it is more rational to consider not the vapor pressure but the specific humidity, q—that is, the mass of water vapor per unit mass of moist air. The amount of water vapor, F, which per second is transported upward through a surface of cross section one square centimeter is, then, $-\mu_e dq/dz$, where μ_e is the eddy diffusivity, which in the air equals the eddy viscosity, and $-dq/dz$ is the vertical gradient of the specific humidity, which is positive when the specific humidity decreases with height. If the vapor pressure, e, is introduced, one obtains approximately

$$F = -\mu_e \frac{0.621}{p} \frac{de}{dz}, \tag{IV, 6}$$

where p is the atmospheric pressure. The heat needed for evaporation at the surface is

$$Q_e = -L_\vartheta \mu_e \frac{0.621}{p} \frac{de}{dz}, \tag{IV, 7}$$

where L_ϑ is the heat of vaporization at the temperature of the surface, ϑ.

The variation of the vapor pressure with increasing height above the sea surface is closely related to the character of the eddy diffusivity. From analogy with experimental results in fluid mechanics it has been concluded that, when the air is in a normal state of turbulence, the eddy diffusivity increases linearly with increasing distance from the sea surface, except within a very thin layer close to the surface. The implication is that, when a steady stage has been reached, the specific humidity must be proportional to the logarithm of the distance from the surface, and observations have to some extent confirmed this conclusion. The reservation that was made when dealing with temperature observations near the sea surface must be made in this case as well, because measurements conducted from vessels are affected by a considerable disturbance of the normal field. It appears, however, that for the proper interpretation of observations it is necessary to assume that next to the sea surface a layer of air of a thickness of a few millimeters exists within which the transport of water vapor takes place by ordinary processes of diffusion. Above this layer, the thickness of which varies with the wind, a transition layer must be assumed, and finally there is a layer within which the eddy diffusivity increases linearly with the distance from the sea surface and also increases with the wind velocity.

The ratio between the amounts of heat given off to the atmosphere as sensible heat and used for evaporation is

$$R = \frac{Q_h}{Q_e} = \frac{c_p}{L_\vartheta} \frac{p}{0.621} \frac{\frac{d\vartheta}{dz}}{\frac{de}{dz}} = 0.64 \frac{p}{1000} \frac{\frac{d\vartheta}{dz}}{\frac{de}{dz}}. \tag{IV, 8}$$

The last expression is obtained by introducing $c_p = 0.240$ and $L_\vartheta = 585$. Thus, the ratio R depends mainly upon the ratio between the temperature and humidity gradients in the air at a short distance from the sea surface. These gradients are difficult to measure but can be replaced approximately by the difference in temperature and vapor pressure at the sea surface and the corresponding values in the air at a height of a few meters:

$$R = 0.64 \frac{p}{1000} \frac{\vartheta_w - \vartheta_a}{e_w - e_a}. \tag{IV, 9}$$

This ratio was derived in a different manner by Bowen, and is often referred to as the "Bowen ratio."

Values of the ratio R can be computed from climatological charts of the oceans, but a comprehensive study has not been made. Calculations based on data contained in the *Atlas of Climatic Charts of the Oceans*, published by the United States Weather Bureau in 1938, show that the ratio varies from one part of the ocean to the other. As a rule, the ratio is small in low latitudes, where it remains nearly constant throughout the year, but is greater in middle latitudes, where it reaches values up to 0.5 in winter and in some areas drops to -0.2 in summer. A negative value indicates that heat is conducted from the atmosphere to the sea. On an average, the value for all oceans appears to lie at about 0.1, meaning that about 10 per cent of the heat surplus that the oceans receive by radiation processes is given off as sensible heat, whereas about 90 per cent is used for evaporation.

There are certain points regarding the character of the evaporation that need to be emphasized. If the water is warmer than the air, the vapor pressure at the sea surface remains greater than that in the air. Under these circumstances, evaporation can always take place and will be greatly facilitated, because, owing to the unstable stratification of the very lowest layers, the turbulence of the air will be fully developed. It must therefore be expected that the greatest evaporation occurs when cold air flows over warm water. If the air is much colder than the water, the air may become saturated with water vapor, and fog or mist may form over the water surfaces. Such fog (*steam fog*) is common in the fall over ponds and small lakes during calm, clear nights. When a wind blows, the moisture will be carried upward, but streaks and columns of fog are often visible over lakes or rivers and are commonly described as "smoke." The process can occasionally be observed near the coast, but it does not occur over the open ocean because the necessarily very great-temperature differences are rapidly eliminated as the distance from the coast increases.

When the sea surface is colder than the air, evaporation can take place only if the air is not saturated with water vapor. In this case, turbulence is reduced and the evaporation must stop when the vapor content of the lowest layer of the atmosphere has reached such a value that the vapor pressure equals that at the sea surface. If warm, moist air passes over a colder sea surface, the direction of transport of water vapor is reversed and condensation takes place on the sea surface in such a way that heat is brought to the surface and not carried away from it. Owing to the fact that this process occurs only when the air is *warmer* than the sea and that then turbulence is greatly reduced, one can expect that condensation of water vapor on the sea will not be of great importance to the heat budget, but it should be borne in mind that this process can, and frequently does, take place when conditions are favorable. In these circumstances, contact with the sea and conduction lower the air temperature to the dew point for a considerable distance above the sea surface. Condensation then takes place in the air and fog is formed. This fog (*advection fog*) is the type that is commonly encountered over the sea. The relation between the frequency of fog or mist and the differences between sea-surface and air temperatures are well illustrated by charts in the *Atlas of Climatic Charts of the Oceans*. As an example, fig. 12 shows the frequency of fog, the difference between the air temperature and the sea-surface temperature, and the prevailing wind direction over the Grand Banks of Newfoundland in March, April, and May. It can be concluded that in spring, when the water is colder than the air, no evaporation takes place in this region, but in the fall and winter, when the water is warmer, evaporation must be great.

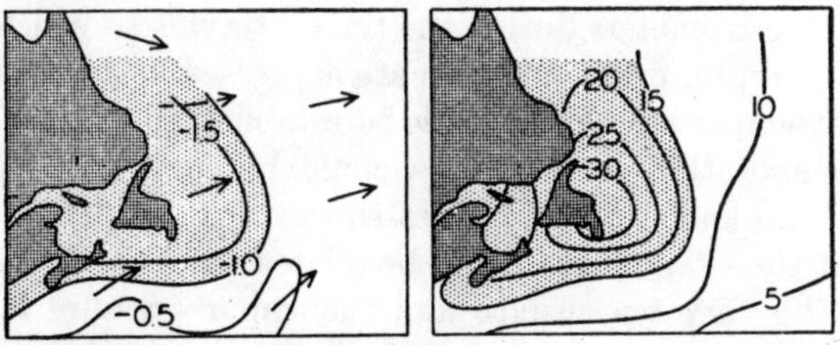

Fig. 12. *Left:* The difference, air temperature minus sea surface temperature, and the prevailing wind directions over the Grand Banks of Newfoundland in March, April, and May. *Right:* Percentage frequency of fog in the same months.

In middle and higher latitudes the sea surface in winter is mostly warmer than the air, and therefore one must expect the evaporation to be at its maximum in winter and not in summer. This conclusion appears contrary to common experience that evaporation from heated water is greater than that from cold water, but there is no contradiction here because greatest evaporation occurs when a water surface is warmer than the air above it, and this is exactly what happens in winter.

Observations and Computations of Evaporation. Present knowledge of the amount of evaporation from the different parts of the oceans is derived partly from observations and partly from computations based on consideration of the heat balance.

Observations have been made by means of pans on board ships, but such observations give too high values of the evaporation from the sea surface, partly because the wind velocity is higher at the level of the pan than at the sea surface, and partly because the difference between the vapor pressure in the air and that of the evaporating surface is greater at the pan than at the sea surface. Analyzing the decrease of the wind velocity and the increase of the vapor pressure between the average level of pans used on shipboard and a level a few centimeters above the sea surface, Wüst arrived at the conclusion that the measured values had to be multiplied by 0.53 in order to represent the evaporation from the sea surface.

In computing the evaporation on the basis of the heat balance, one has to start out from equation (IV, 2) (p. 50). Introducing the ratio, $R = Q_h/Q_e$, putting $Q_s - Q_b = Q_r$, and taking into account that the evaporation, E, is obtained in centimeters by dividing Q_e by the latent heat of vaporization, L, one obtains

$$E = \frac{Q_r - Q_v - Q_\vartheta}{L(1 + R)}. \qquad \text{(IV, 10)}$$

In this form the equation representing the heat balance has found wide application for computation of evaporation. The result gives the evaporation in centimeters during the time intervals to which the values Q_r, and so on, apply, provided these are expressed in gram calories.

Several attempts have been made to establish empirical formulas by which the evaporation from the sea could be computed from meteorological data and knowledge of the sea-surface temperature. It has been assumed that the evaporation could be related to the difference in vapor pressure at the very sea surface and the vapor pressure in the air as observed on shipboard, $e_w - e_a$, and to the wind velocity, W_a. Various formulas of the type $E = f[(e_w - e_a)W_a]$ have been proposed, but no consistent results have been obtained because each formula has been developed in order to represent some specific set of very uncertain observations. Another approach was suggested by Sverdrup, who, on the basis of results in fluid mechanics as to the turbulence of the air over a rough surface, established a formula for the evaporation, using, in part, constants that had been determined by laboratory experiments, and, in part, constants that were obtained from the character of the variation of vapor pressure with increasing height above the sea surface. Somewhat similar but more complicated formulas have been derived by Millar and by Montgomery.

These investigations indicate that at wind velocities below 4 to 5 m/sec (10 miles per hour) the evaporation is relatively small, partly because at low wind velocities and stable stratification of the air the sea surface has the character of a hydrodynamically smooth surface (p. 120), and partly because at higher wind velocities the evaporation is greatly increased on account of spray. Furthermore, it is found that at moderate and high wind velocities the evaporation can be written

$$E = k(e_w - e_a)W_a, \qquad \text{(IV, 11)}$$

where the factor k is nearly a constant, the numerical value of which may be determined when the character of the variation of vapor content with height is better known. Introducing average annual values of the vapor pressure (in millibars) and average annual wind velocity (in meters per second), one obtains the evaporation in centimeters per year by putting $k = 3.6$, but this numerical value depends upon the heights at which the vapor pressure in the air and the wind velocity are measured, and should therefore be used with caution.

Average Annual Evaporation from the Oceans. On the basis of pan measurements conducted in different parts of the ocean, Wüst found that the average evaporation from all oceans amounts to 93 cm per year, and he considers this value correct to within 10 or 15 per cent. W. Schmidt computed the evaporation for E by means of equation (IV, 10), in which the terms Q_v and Q_ϑ can be omitted when considering the oceans as a whole. Schmidt introduced a high value of R, and on the basis of the available data as to incoming radiation and back radiation he found a total evaporation of 76 cm a year. A revision based on more recent measurements of radiation (Mosby) and use of $R = 0.1$ resulted in a value of 106 cm a year. The latter value represents an upper limit and may be 10 to 15 per cent too high, wherefore it appears that Wüst's result is nearly correct.

It is of interest in this connection to give some figures regarding the relation between evaporation and precipitation over the oceans, the land areas, and the whole earth. According to Wüst the total evaporation from the oceans amounts to 334,000 km^3/year, of which 297,000 km^3 returns to the sea in the form of precipitation, and the difference, 37,000 km^3, must be supplied by run-off. The total amount of precipitation falling on the land is 99,000 km^3, of which amount a little over one third, 37,000 km^3, is supplied by evaporation from the oceans and 62,000 km^3 is supplied by evaporation from inland water areas or directly from the moist soil. For the sake of comparison it may be mentioned that the maximum capacity of Lake Mead above Boulder Dam is 45 km^3.

Evaporation in Different Latitudes. From pan observations at sea, Wüst has derived average values of the evaporation from the different oceans in different latitudes (table 10, p. 72). By means of the energy equation, one can compute similar annual values, assuming that the net transport of heat by ocean currents can be neglected. Such a computation has been carried out for the Atlantic Ocean, making use of Kimball's data as to the incoming radiation of the observed temperatures and humidities for determining effective back radiation. In fig. 13 are shown Wüst's values for the annual evaporation between latitudes 50° N and 50° S in the Atlantic Ocean and the corresponding values as derived from the energy equation. The low evaporation in the equatorial regions which both methods show can be ascribed to the higher relative humidities and the lower wind velocities of that area, if one considers the processes of evaporation, or it can be ascribed to the effect of the prevailing cloudiness, if one considers the energy relations. The great evaporations in the areas of the strong trade winds appear clearly, but in the Southern Hemisphere the observations give the highest values of the evaporation nearer to the Equator than do the computations. The discrepancy may be due to the fact that in the course of a year the wind systems change their distances from the Equator and that the observations have not been distributed evenly over the year. The energy equation has also been used by McEwen for computing values of evaporation over the eastern Pacific Ocean between latitudes 20° N and 50° N. His figures agree with those obtained by Wüst for the same latitudes.

It appears that the average annual values of the evaporation in different latitudes are well established, but the evaporation also varies from the eastern to the western parts of the oceans and with the seasons.

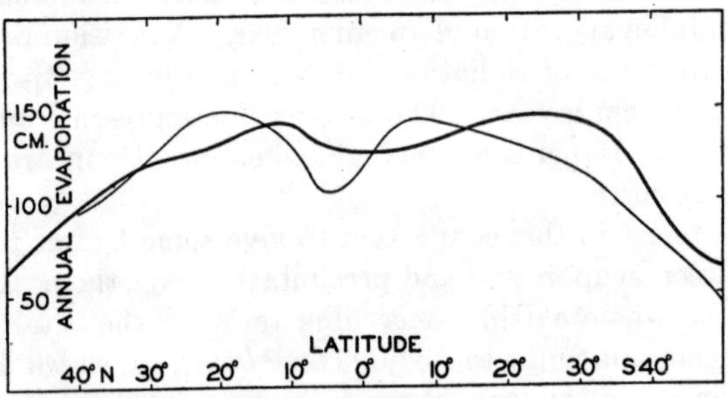

Fig. 13. Annual evaporation from the Atlantic Ocean between lat. 50°N and 50°S. The thin curve is based on observations (Wüst) and the heavy curve on computations, using the energy equation.

These variations, which are closely related to the ocean currents, are of great importance to the circulation of the atmosphere, because the supply of water vapor which later on condenses and gives off its latent heat represents a large portion of the supply of energy. Approximate values of the evaporation from different parts of the oceans and in different seasons can be found by means of the method proposed by Sverdrup, and will be dealt with after the ocean currents have been discussed.

Annual Variation of Evaporation. The character of the annual variation of evaporation can be examined by means of the energy equation

$$q_a = q_e(1 + R) = q_r - q_v - q_\vartheta. \qquad \text{(IV, 12)}$$

The quantity q_ϑ can be computed if the annual variation of temperature due to processes of heating and cooling is known at all depths where such annual variations occur. The annual variation of temperature at the surface has been examined, but only few data are available from subsurface depths, the most reliable being those that have been compiled by Helland-Hansen from an area in the eastern North Atlantic with its center in 47° N and 12° W. The radiation income in that area can be obtained from Kimball's data, and the back radiation can be found by means of the diagram in fig. 11; in this region the transport by currents, q_v, can be neglected. In fig. 14A are represented the annual variation of the net surplus of radiation, q_r, the annual variation of the amount of heat used for changing the temperature of the water, q_ϑ, and the difference between these two amounts, q_a, which represents the total amount of heat given off to the atmosphere. The greater part of the last amount is used for evaporation, and therefore the curve marked q_a represents approximately the annual variation of the evaporation, which shows a maximum in the fall and early winter, a secondary minimum in February, followed by a secondary maximum in March, and a low minimum in summer. In June and July no evaporation takes place.

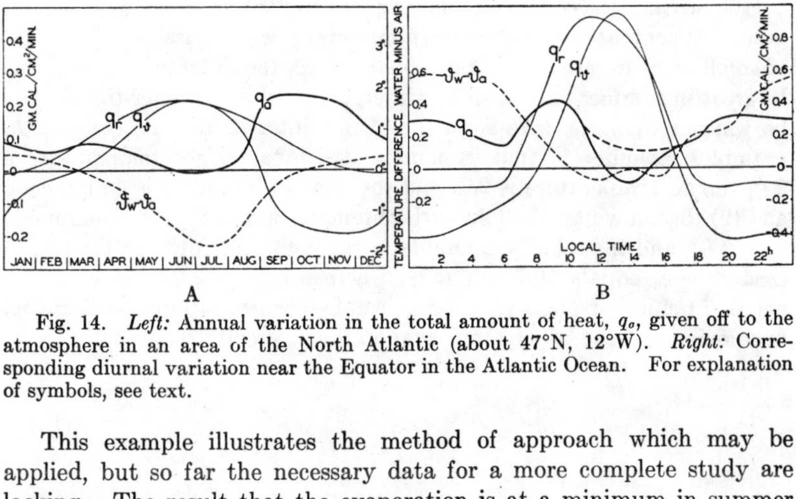

Fig. 14. *Left:* Annual variation in the total amount of heat, q_a, given off to the atmosphere in an area of the North Atlantic (about 47°N, 12°W). *Right:* Corresponding diurnal variation near the Equator in the Atlantic Ocean. For explanation of symbols, see text.

This example illustrates the method of approach which may be applied, but so far the necessary data for a more complete study are lacking. The result that the evaporation is at a minimum in summer and at a maximum in fall and early winter is in agreement with the conclusions that were drawn when discussing the process of evaporation in general.

Diurnal Variation of Evaporation. The diurnal variation of evaporation can be examined in a similar manner, but at the present time suitable data are available only at four *Meteor* stations near the Equator in the Atlantic Ocean. In fig. 14B the curves marked q_r and q_ϑ correspond to the similar curves in fig. 14A, and the difference between these, q_a, shows the amount of heat lost during 24 hours, which is approximately proportional to the evaporation. The diurnal variation of evaporation in the Tropics appears to have considerable similarity to the annual variation in middle latitudes, and is characterized by a double period with maxima in the late forenoon and the early part of the night and minima at sunrise and in the early afternoon hours. It is possible that the afternoon minimum appears exaggerated, owing to uncertainties as to the absolute values of q_r and q_ϑ. The total diurnal evaporation was 0.5 cm, but the sky was nearly clear on the four days that were examined, and the average diurnal value is therefore smaller. The double diurnal period of evaporation appears to be characteristic of the Tropics, but in middle latitudes a single period with maximum values during the night probably dominates.

It may be added here that the annual variation of evaporation from small inland lakes is quite different from that from the oceans, in general reaching maximum values in summer. There are many reasons for this difference. In the first place, the annual range of the lake-surface temperature is great, the temperature varying between, say, 25° and 0°. In the second place, the air blowing over a lake is often dry, and its humidity content will not increase materially when passing over a lake of small or moderate size. As a consequence, the difference $e_w - e_a$ will be great in summer and small in winter, although in summer the air may be warmer than the lake surface and in winter it may be colder. As examples, assume (1) that in summer the lake surface temperature is 20°, the air temperature is 25°, and the relative humidity is 40 per cent, and (2) that in winter the lake surface temperature is 5°, the air temperature is 0°, and the relative humidity is again 40 per cent. In the former case $e_w - e_a$ equals 10.7 millibars, whereas in the latter case $e_w - e_a$ equals 7.6 millibars, and at the same wind velocity evaporation is greater in summer. If the lake freezes, evaporation is greatly reduced.

CHAPTER V

General Distribution of Salinity, Temperature, and Density

Salinity of the Surface Layer

SURFACE SALINITY. In all oceans the surface salinity varies with latitude in a similar manner. It is at a minimum near the Equator, reaches a maximum in about latitudes 20°N and 20°S, and again decreases toward high latitudes.

Table 10 contains average values of the surface salinity, the evaporation, and the precipitation, as well as the difference between the last

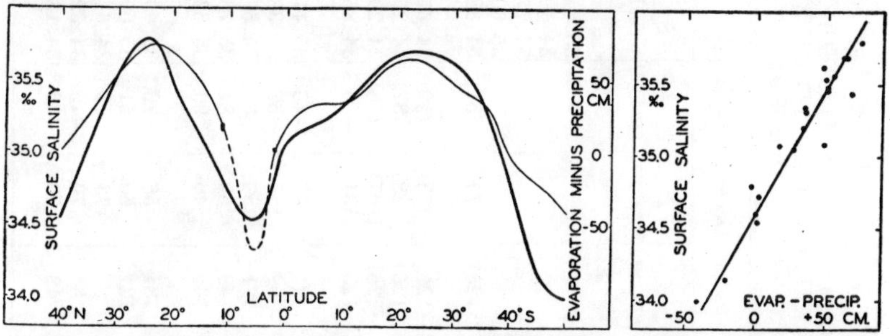

Fig. 15. *Left:* Average values of surface salinity for all oceans, and the difference, evaporation minus precipitation, plotted against latitude. *Right:* Corresponding values of surface salinity, and the difference, evaporation minus precipitation, plotted against each other (according to Wüst).

quantities, for the three major oceans and for all combined, according to Wüst. On the basis of these values, Wüst has shown that for each ocean the deviation of the surface salinity from a standard value is directly proportional to the difference between evaporation, E, and precipitation, P. In fig. 15 are plotted the surface salinities for all oceans and the difference $E - P$ (in centimeters per year), as functions of latitude, and the corresponding values of salinity and the difference, $E - P$, are plotted against each other. Omitting the values at 5°N because they disagree with all others, the values fall nearly on a straight line leading to the empirical relationship

$$S = 34.60 + 0.0175(E - P). \qquad (V, 1)$$

TABLE 10

AVERAGE VALUES OF SALINITY, S, EVAPORATION, E, AND PRECIPITATION, P, AND THE DIFFERENCE, $E - P$, FOR EVERY FIFTH PARALLEL OF LATITUDE BETWEEN 40°N AND 50°S
(After Wüst)

Latitude	Atlantic Ocean				Indian Ocean				Pacific Ocean				All Oceans			
	S (°/₀₀)	E (cm/yr)	P (cm/yr)	$E - P$ (cm/yr)	S (°/₀₀)	E (cm/yr)	P (cm/yr)	$E - P$ (cm/yr)	S (°/₀₀)	E (cm/yr)	P (cm/yr)	$E - P$ (cm/yr)	S (°/₀₀)	E (cm/yr)	P (cm/yr)	$E - P$ (cm/yr)
40°N	35.80	94	76	18					33.64	94	93	1	34.54	94	93	1
35	36.46	107	64	43					34.10	106	79	27	35.05	106	79	27
30	36.79	121	54	67					34.77	116	65	51	35.56	120	65	55
25	36.87	140	42	98					35.00	127	55	72	35.79	129	55	74
20	36.47	149	40	110	(35.05)	(125)	(74)	(51)	34.88	130	62	68	35.44	133	65	68
15	35.92	145	62	83	(35.07)	(125)	(73)	(52)	34.67	128	82	46	35.09	130	82	48
10	35.62	132	101	31	(34.92)	(125)	(88)	(37)	34.29	123	127	− 4	34.72	129	127	2
5	34.98	105	144	−39	(34.82)	(125)	(107)	(18)	34.29	102	(177)	(−75)	34.54	110	177	−67
0	35.67	116	96	20	35.14	125	131	− 6	34.85	116	98	18	35.08	119	102	17
5°S	35.77	141	42	99	34.93	121	167	−46	35.11	131	91	40	35.20	124	91	33
10	36.45	143	22	121	34.57	99	156	−57	35.38	131	96	35	35.34	130	96	34
15	36.79	138	19	119	34.75	121	83	38	35.57	125	85	40	35.54	134	85	49
20	36.54	132	30	102	35.15	143	59	84	35.70	121	70	51	35.69	134	70	64
25	36.20	124	40	84	35.45	145	46	99	35.62	116	61	55	35.69	124	62	62
30	35.72	116	45	71	35.89	134	58	76	35.40	110	64	46	35.62	111	64	47
35	35.35	99	55	44	35.60	121	60	61	35.00	97	64	33	35.32	99	64	35
40	34.65	81	72	9	35.10	83	73	10	34.61	81	84	− 3	34.79	81	84	− 3
45	34.19	64	73	− 9	34.25	64	79	−15	34.32	64	85	−21	34.14	64	85	−21
50	33.94	43	72	−29	33.87	43	79	−36	34.16	43	84	−41	33.99	43	84	−41

Wüst points out that such an empirical relationship is found because the surface salinity is mainly determined by three processes: decrease of salinity by precipitation, increase of salinity by evaporation, and change of salinity by processes of mixing. If the surface waters are mixed with water of a constant salinity, and if this constant salinity is denoted by S_0, the change of salinity due to mixing must be proportional to $S_0 - S$, where S is the surface salinity. The change of salinity due to processes of evaporation and precipitation must be proportional to the difference, $E - P$. The local change of the surface salinity must be zero; that is,

$$\frac{\partial S}{\partial t} = a(S - S_0) + b(E - P) = 0,$$

or

$$S = S_0 + k(E - P). \qquad (V, 2)$$

As this simple formula has been established empirically, it must be concluded that the surface water is generally mixed with water of a salinity that is, on an average, 34.6°/₀₀. This value represents approximately the average value of the salinity at a depth of 400 to 600 m, and it appears therefore that vertical mixing is of great importance to the general distribution of surface salinity. This concept is confirmed by the fact that the standard value of the salinity differs for the different oceans. For the North Atlantic and the North Pacific, Wüst obtains similar relationships, but the constant term, S_0, has the value 35.30°/₀₀ in the North Atlantic Ocean and 33.70°/₀₀ in the North Pacific Ocean. The corresponding average values of the salinity at a depth of 600 m are 35.5°/₀₀ and 34.0°/₀₀, respectively. For the South Atlantic and the South Pacific Oceans, Wüst finds that S_0 = 34.50°/₀₀ and 34.64°/₀₀, respectively, and that the average salinity at 600 m in both oceans is about 34.5°/₀₀. In these considerations the effect of ocean currents on the distribution of surface salinity has been neglected, and the simple relations obtained indicate that transport by ocean currents is of minor importance as far as average conditions are concerned. The difference between evaporation and precipitation, $E - P$, is, on the other hand, of primary importance, and, because this difference is closely related to the circulation of the atmosphere, one is led to the conclusion that the average values of the surface salinity are, to a great extent, controlled by the character of the atmospheric circulation.

The distribution of surface salinity of the different oceans is shown in chart 3, in which the general features that have been discussed are recognized, but the details are so closely related to the manner in which the water masses are formed and to the types of currents that they cannot be dealt with here.

Periodic Variations of the Surface Salinity. Over a large area, variations in surface salinity depend mainly upon variations in the difference between evaporation and precipitation. From Böhnecke's monthly charts of the surface salinity of the North Atlantic Ocean, mean monthly values have been computed for an area extending between latitudes 18° and 42° N, omitting the coastal areas in order to avoid complications due to shifts of coastal currents. The results of this computation show the highest average surface salinity, 36.70°/₀₀, in March and the lowest, 36.59°/₀₀, in November. The variations from one month to another are irregular, but on the whole the salinity is somewhat higher in spring than it is in the fall.

Harmonic analysis leads to the result

$$S = 36.641 + 0.021 \cos\left(2\pi \frac{t}{T} - 80°\right), \qquad \text{(V, 3)}$$

and thus,

$$\frac{\partial S}{\partial t} = 0.021 \frac{2\pi}{T} \cos\left(2\pi \frac{t}{T} - 350°\right). \qquad \text{(V, 4)}$$

According to (V, 2), $\partial S/\partial t$ is proportional to $E - P$, and it follows therefore that the excess of evaporation over precipitation is at a minimum at the end of June and at a maximum at the end of December. This annual variation corresponds closely to the annual variation of evaporation (p. 68), for which reason it appears that in the area under consideration the annual variation of the surface salinity is mainly controlled by the variation in evaporation during the year. For a more exact examination, subsurface data are needed, but nothing is known as to the annual variation of salinity at subsurface depths.

From the open ocean, no data are available as to the diurnal variation of the salinity of the surface waters, but it may be safely assumed that such a variation is small, because neither the precipitation nor the evaporation can be expected to show any considerable diurnal variation.

Temperature of the Surface Layers

Surface Temperature. The general distribution of surface temperature cannot be treated in the same manner that Wüst employed for the salinity, because the factors controlling the surface temperature are far more complicated. The discussion must be confined to presentation of empirical data and a few general remarks.

Table 11 contains the average temperatures of the oceans in different latitudes, according to Krümmel, except in the case of the Atlantic Ocean, for which new data have been compiled by Böhnecke. In all oceans the highest values of the surface temperature are found somewhat to the north of the Equator, and this feature is probably related to the character of the atmospheric circulation in the two hemispheres. The region of the highest temperature, the thermal Equator, shifts with the season, but only in a few areas is it displaced to the Southern Hemisphere at any season. The larger displacements are all in regions in which the surface currents change during the year because of changes in the prevailing winds, and this feature also is therefore closely associated with the character of the atmospheric circulation. The surface temperatures in the Southern Hemisphere are generally somewhat lower than those in the Northern Hemisphere, and again the difference can be ascribed to difference in the character of the prevailing winds, and perhaps also to a wide-spread effect of the cold, glacier-covered Antarctic Continent.

TABLE 11

AVERAGE SURFACE TEMPERATURE OF THE OCEANS BETWEEN PARALLELS OF LATITUDE

North latitude	Atlantic Ocean	Indian Ocean	Pacific Ocean	South latitude	Atlantic Ocean	Indian Ocean	Pacific Ocean
70°–60°	5.60	.....	.....	70°–60°	− 1.30	− 1.50	− 1.30
60 –50	8.66	.....	5.74	60 –50	1.76	1.63	5.00
50 –40	13.16	.....	9.99	50 –40	8.68	8.67	11.16
40 –30	20.40	.....	18.62	40 –30	16.90	17.00	16.98
30 –20	24.16	26.14	23.38	30 –20	21.20	22.53	21.53
20 –10	25.81	27.23	26.42	20 –10	23.16	25.85	25.11
10 – 0	26.66	27.88	27.20	10 – 0	25.18	27.41	26.01

The average distribution of the surface temperature of the oceans in February and August is shown in charts 1 and 2. Again the distribution is so closely related to the formation of the different water masses and the character of the currents that a discussion of the details must be postponed.

DIFFERENCE BETWEEN AIR- AND SEA-SURFACE TEMPERATURES. It was pointed out that in all latitudes the ice-free oceans received a surplus of radiation, and that therefore in all latitudes heat is given off from the ocean to the atmosphere in the form of sensible heat or latent heat of water vapor. The sea-surface temperature must therefore, on an average, be higher than the air temperature. Observations at sea have shown that such is the case, and from careful determinations of air temperatures over the oceans it has furthermore been concluded that the difference, air- minus sea-surface temperature, is greater than that derived from routine ships' observations. In order to obtain an exact value, it is necessary to measure the air temperature on the windward side of the vessel at a locality where no eddies prevail and where the air reaches the thermometer without having passed over any part of the vessel. For measurements of the temperature a ventilated thermometer must be used. According to the *Meteor* observations the air temperature over the South Atlantic Ocean between latitudes 55°S and 20°N is, on an average, 0.8° lower than that of the surface, whereas in the same region the Netherlands Atlas of air- and sea-temperatures gives an average difference of only 0.1 degree. The reason for this discrepancy is that, on an average, the air temperatures as determined on commercial vessels are about 0.7° too high, owing to the ships' heat. The *Meteor* result as to the average value of the difference, $\vartheta_w - \vartheta_a$, is in good agreement with results obtained on other expeditions that took special precautions for obtaining correct air temperatures. Present atlases of air- and sea-surface temperatures have been prepared from the directly observed values on board commercial vessels without application of a correction to the air temperatures. This correction is so small that it is of minor importance when the atlases are used for climatological studies, but, in any studies which require knowledge of the exact difference between air and surface temperatures, it is necessary to be aware of the systematic error in the air temperature.

The difference of 0.8° between air and surface temperatures, as derived from the *Meteor* observations, is based on measurements of air temperature at a height of 8 m above sea level. At the very sea surface the air temperature must coincide with that of the water, and consequently the air temperature decreases within the layers directly above the sea. The most rapid decrease takes place, however, very close to the sea surface, and at distances greater than a few meters the decrease is so slow that it is immaterial whether the temperature is measured at 6, 8, or 10 m above the surface. The height at which the air temperature has been observed on board a ship exercises therefore a minor influence upon the accuracy of the result, and discrepancies due to differences in the height of observations are negligible compared to the errors due to inadequate exposure of the thermometer.

The statement that the air temperature is lower than the water temperature is correct only when dealing with average conditions. In middle latitudes the difference, $\vartheta_w - \vartheta_a$, generally varies during the year in such a manner that in winter the air temperature is lower than the sea-surface temperature, whereas in summer the difference is reduced or the sign is reversed. The difference also varies from one region to another according to the character of the circulation of the atmosphere and of the ocean currents. These variations are of great importance to the local heat budget of the sea, because the exchange of heat and vapor between the atmosphere and the ocean depends greatly upon the temperature difference.

It was shown that in middle latitudes the amount of heat given off from the ocean to the atmosphere is, in general, great in winter and probably negligible in summer. Owing to this annual variation in the heat exchange, one must expect the air over the oceans in winter to be much warmer than the air over the continents, but in summer the reverse conditions should be expected. That such is the case is evident from a computation of the average temperature of the air between latitudes 20°N and 80°N along the meridian of 120°E, which runs entirely over land, and along the meridian of 20°W, which runs entirely over the ocean. In January the average temperature along the "land meridian" of 120°E is −15.9°, but along the "water meridian" of 20°W it is 6.3°. In July the corresponding values are 19.4° and 14.6°, respectively. Thus in January the air temperature between 20°N and 80°N over the water meridian is 22.2° higher than that over the land meridian, whereas in July it is 4.8° lower. The mean annual temperature is 7.0° higher along the water meridian.

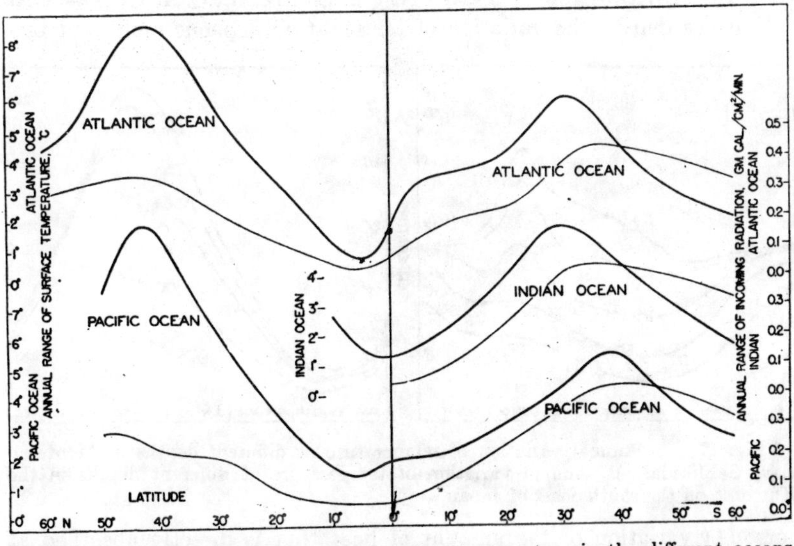

Fig. 16. Average annual ranges of surface temperature in the different oceans plotted against latitude (heavy curves) and the corresponding ranges in the radiation income (light curves).

Annual Variation of Surface Temperature. The annual variation of surface temperature in any region depends upon a number of factors, foremost among which are the variation during the year of the radiation income and the character of the ocean currents and of the prevailing winds. The character of the annual variation of the surface temperature changes from one locality to another, but a few of the general features can be pointed out. The heavy curves in fig. 16 show the average range of the surface temperature in different latitudes in the Atlantic, the Indian, and the Pacific Oceans. The range represents the difference between the average temperatures in February and August, and the thin lines in the same figure show the range of the radiation income. The curves bring out two characteristic features. In the first place, they show that the annual range of surface temperature is much greater in the North Atlantic and the North Pacific Oceans than in the southern oceans. In the second place, they show that in the southern oceans the temperature range is definitely related to the range in radiation income, whereas in the northern oceans no such definite relation appears to exist. The great ranges in the northern oceans are associated with the character of the prevailing winds and, particularly, with the fact that cold winds blow from the continents toward the ocean and greatly reduce the winter temperatures. Near the Equator a semiannual variation is present, corresponding to the semiannual period of radiation income, but in middle and higher latitudes the annual period dominates.

Annual Variation of Temperature in the Surface Layers. At subsurface depths the variation of temperature depends upon four fac-

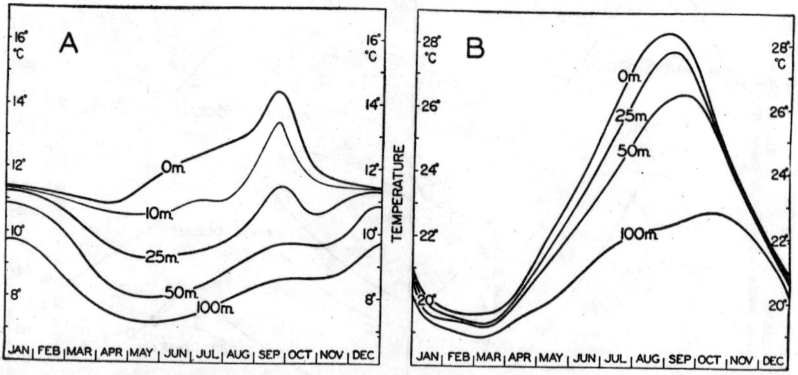

Fig. 17. A. Annual variation of temperature at different depths in Monterey Bay, California. B. Annual variation of temperature at different depths in the Kuroshio off the south coast of Japan.

tors: (1) variation of the amount of heat that is directly absorbed at different depths, (2) the effect of heat conduction, (3) variation in the currents related to lateral displacement of water masses, and (4) the effect of vertical motion. The annual variation of temperature of subsurface depths cannot be dealt with in a general manner, owing to lack of data, but it is again possible to point out some outstanding characteristics, using two examples from the Pacific and one from the Atlantic Oceans. The effects of all four of the important factors are illustrated in fig. 17A, which shows the annual variation of temperature at the surface and at depths of 25, 50, and 100 m at Monterey Bay, California, according to Skogsberg. Skogsberg divides the year into three periods: the period of the Davidson Current, lasting from the middle of November to the middle of February, the period of upwelling, between the middle of February and the end of July, and the oceanic period, from the end of July to the middle of November. The California Current off Monterey Bay, during the greater part of the year, is directed to the south, but during winter, from the middle of November to the middle of February, an inshore flow to the north, the Davidson Current, is present (p. 204). The water of this inshore flow is characterized by relatively high and uniform temperature, and appears in the annual variation of temperature as warm water at subsurface depths. The upper homogeneous layer is relatively thick, as is evident from the fact that the temperature is nearly the same at 25 m as it is at the surface, and that at 50 m it is only slightly lower. At the end of February the California Current again reaches to the coast, and under the influence of the prevailing northwesterly winds an overturn of the upper layers occurs which is generally described as upwelling (p. 131). During the period of upwelling, vertical motion near the coast brings water of relatively low temperature toward the surface. Consequently, the temperatures at given depths decrease when the upwelling begins. This decrease is brought out in fig. 17A by the downward trend of the temperature at 25, 50, and 100 m, at which depths the minimum temperature is reached at the end of May. The much higher temperature at the surface as compared to that at 25 m shows that a thin surface layer is subject to heating by radiation, and from the variation of temperature at 10 m, which is shown by a thin line, it is evident that the effect of heating is limited to the upper 10 m. As the upwelling gradually ceases toward the end of August, a sharp rise in temperature takes place both at the surface and at subsurface depths, and the peaks shown by the temperature curves in September are results of heating and conduction. Thus, the annual march of temperature can be attributed to changes in currents, to vertical motion associated with upwelling, to seasonal heating and cooling, and to heat conduction.

The annual variation of temperature in the Kuroshio, off the south coast of Japan, as shown in fig. 17B, gives an entirely different picture. The annual variation has the same character at all depths between the surface and 100 m, with a minimum in late winter and a maximum in late summer or early fall, but the range of the variation decreases with depth, and the time of maximum temperature occurs later with increasing depth. From the course of the curves, it may be concluded that the annual variation is due to heating and cooling near the surface and is transmitted to greater depths by processes of conduction. This assumption appears to be correct, but the heating and cooling also depend on excessive cooling in winter by cold and dry winds blowing toward the sea, and are caused only partly by variations in the net radiation.

In order to be certain that observed temperature variations are related to processes of heating and cooling only, it is necessary to determine whether the water in a given locality is of the same character throughout the year. For this purpose, Helland-Hansen developed a method that is applicable in areas in which it is possible to establish a definite relation between temperature and salinity (p. 86). He assumed that any temperature value above or below that determined by the temperature-salinity relation could be considered as resulting from heating or cooling of the water, and used the method within three areas in the eastern North Atlantic. Fig. 18 shows the curves that he determined for an area off the Bay of Biscay with its center in approximately 47°N and 12°W. The character of the curves, the reduction of the range, and the displacement of the times at which maxima occurred clearly show that one has to deal with heat conduction. In this case the variation in the heat content corresponds nearly to the variation in net radiation, whereas in the Kuroshio the additional effect of excessive cooling in winter by winds from the continent leads to much greater variations in temperature and heat content.

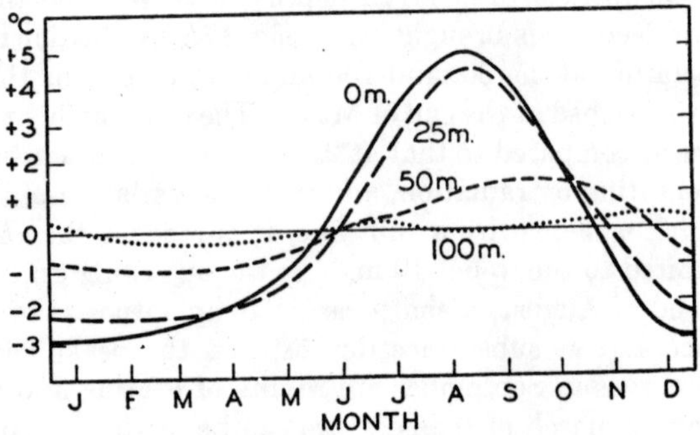

Fig. 18. Annual variation of temperature at different depths off the Bay of Biscay in about 47°N and 12°W.

These examples serve to illustrate different types of annual variation of temperature that may be encountered in different localities and also to stress that conclusions as to the temperature variations associated with processes of local heating are valid only if the data are such that the influences of shifting currents and vertical motion can be eliminated.

Diurnal Variation of Surface Temperature. The range of the diurnal variation of surface temperature of the sea is, on an average, not more than 0.2 to 0.3°. Earlier observations gave somewhat higher values, particularly in the Tropics, but new, more careful measurements and re-examination of earlier data in which doubtful observations were eliminated have shown that the range of diurnal variation is quite small. It can be stated that in general the diurnal variation of surface temperature in lower latitudes can be represented by a sine curve with extreme values between 2:30^h and 3^h and between 14:30^h and 15^h, and a range of 0.3° to 0.4°. In higher latitudes the extreme values come later and the range is even smaller. In the Tropics the *Meteor* observations give ranges of only 0.2° to 0.3°. The *Meteor* and the *Challenger* data show that close to the Equator the diurnal variation of surface temperature is somewhat unsymmetrical, the temperature increasing rapidly after sunrise and decreasing slowly after sunset, but at greater distances from the Equator the curve becomes somewhat more symmetrical.

In some coastal areas the changes through the year of the range of diurnal variation of surface temperature have been examined. At 44 stations around the British Isles, Dickson found that on an average the diurnal range varied between 0.20° in December and 0.69° in May. At individual stations, both the mean annual range and the variation of the range from month to month were dependent upon the exposure of the locality and the depth of the water at which the measurements were made. Similar results were derived by van der Stock from observations at the Netherlands light ship on Schouwenbank, where during the year the range varied from about 0.15° in January to 0.44° in June. This annual variation in range is closely related to the annual variation in the diurnal amount of net heat received by processes of radiation.

The range of the diurnal variation of temperature depends upon the cloudiness and the wind velocity. From observations in the Tropics, Schott found the mean and extreme values that are shown in table 12.

TABLE 12

RANGE OF DIURNAL VARIATION OF SURFACE TEMPERATURE IN THE TROPICS

Wind and cloudiness	Temperature range, °C		
	Average	Maximum	Minimum
1. Moderate to fresh breeze			
a. Sky overcast	0.39	0.6	0.0
b. Sky clear	0.71	1.1	0.3
2. Calm or very light breeze			
a. Sky overcast	0.93	1.4	0.6
b. Sky clear	1.59	1.9	1.2

Similar results but higher numerical values were found by Wegemann from the *Challenger* data. In both cases the numerical values may be somewhat in error, but the character of the influence of cloudiness and wind is quite evident. With a clear sky the range of the diurnal variation is great, but with great cloudiness it is small; at calm or light breeze it is great, and at moderate or high wind it is small. The effect of cloudiness is explained by the decrease of the diurnal amplitude of the incoming radiation with increasing cloudiness. The effect of the wind is somewhat more complicated, but the main feature is that at high wind velocities the wave motion produces a thorough mixing in the surface layers and the heat that is absorbed in the upper meters is distributed over a thick layer, leading to a small range of the temperature, whereas in calm weather a corresponding intensive mixing does not take place, the heat is not distributed over a thick layer, and, consequently, the range of the temperature near the surface is much greater.

Diurnal Variation of Temperature in the Upper Layers. Knowledge as to the diurnal variation of temperature at depths below the surface is very scanty. It can be assumed that the depth at which the diurnal variation is perceptible will depend greatly upon the stratification of the water. A layer of sharp increase of density a short distance below the free water surface will limit the conduction of heat to such an extent that diurnal variation of temperature will be present above the boundary layer only.

On the *Meteor* expedition, hourly temperature observations were made at the surface and at a depth of 50 m at a few stations in the Tropics where an upper homogeneous layer was present which had a thickness of 70 m. Defant has shown that in these cases the diurnal oscillation of temperature at subsurface depths is in agreement with the laws that have been derived on the assumption of a constant heat conductivity. At 50 m the amplitude of the diurnal variation was reduced to less than two tenths of the amplitude at the surface, and the maximum occurred about 6.5 hours later.

The diurnal variation of sea temperature is, in general, so small that it is of little importance to the physical and biological processes in the sea, but knowledge of the small variations is essential to the study of the diurnal exchange of heat between the atmosphere and the sea (p. 69). The data that are available for this purpose, however, are very inadequate at the present time.

Distribution of Density

The distribution of the density of the ocean waters is related to two features. In a vertical direction the stratification is generally stable (p. 87 and p. 100), and in a horizontal direction differences in density can exist only in the presence of currents. The general distribution of density is therefore closely related to the character of the currents, but for the present purpose it is sufficient to emphasize the fact that in every ocean region water of a certain density that sinks from the sea surface tends to sink to and spread at depths where that density is found.

Since the density of sea water depends on its temperature and salinity, all processes that alter the temperature or the salinity influence the density. At the surface the density will be decreased by heating, addition of precipitation, melt-water from ice, or by run-off from land, and will be increased by cooling, evaporation, or formation of ice. If the density of the surface water is increased beyond that of the underlying strata, vertical convection currents arise that lead to the formation of a layer of homogeneous water. Where intensive cooling, evaporation, or freezing takes place, the vertical convection currents penetrate to greater and greater depths until the density has attained a uniform value from the surface to the bottom. When this state has been established, continued increase of the density of the surface water leads to an accumulation of the densest water near the bottom, and, if the process goes on in an area that is in free communication with other areas, this bottom water of great density spreads to other regions. If the increase of density at the surface stops before the convection currents have reached the bottom, the process may lead to the formation of water that spreads at an intermediate level.

In the open oceans the temperature of the surface water in lower and middle latitudes is so high that the density of the water remains low, even in regions where high salinities occur through excess of evaporation. In these latitudes, convection currents are limited to a relatively thin layer near the surface and do not lead to formation of deep or bottom water. Such formation takes place mainly in high latitudes, where, however, the excess of precipitation in most regions prevents the development of convection currents that reach to great depths. This excess of precipitation is so great that deep and bottom water is formed only in two cases: (1) if water of high salinity has been carried into high latitudes by currents and is cooled, and (2) if water of relatively high salinity freezes.

The first conditions are encountered in the North Atlantic Ocean, where water of the Gulf Stream system, the salinity of which has been raised in lower latitudes by excessive evaporation, is carried into high latitudes. In the Irminger Sea, between Iceland and Greenland, and in the Labrador Sea, this water is partly mixed with cold water of low salinity that flows out from the Polar Sea (p. 156). This mixed water has a relatively high salinity, and, when it is cooled in winter, convection currents that may reach from the surface to the bottom develop before any freezing of ice begins. In this manner, deep and bottom water is formed that has a high salinity and a temperature that lies several degrees above the freezing point of the water (table 19, p. 170). A similar process takes place in the Norwegian Sea, but there deep and bottom water is formed at a temperature which deviates only slightly from the freezing temperature (p. 157).

In the Arctic the second process is of minor importance. There, in the regions where freezing occurs, the salinity of the surface layers is very low, mainly owing to the enormous masses of fresh water that are carried into the sea by the Siberian rivers. Close to the Antarctic continent, however, formation of bottom water by freezing is of the greatest importance. At some distance from the Antarctic continent the great excess of precipitation maintains a low surface salinity, and in these areas winter freezing is not great enough to increase the salinity sufficiently to form bottom water, but on some parts of the continental shelf surrounding Antarctica rapid freezing in winter produces homogeneous water that attains a higher density than the water off the shelf, and therefore flows down the continental slope to the greatest depths. When sinking, the water is mixed with circumpolar water of somewhat higher temperature and salinity, and therefore the resulting bottom water has a temperature slightly above freezing point (p. 155). An active production of bottom water takes place to the south of the Atlantic Ocean, but not within the Antarctic part of the Pacific Ocean.

In some isolated adjacent seas the evaporation may be so intense that a moderate cooling leads to the formation of bottom water. This is the case in the Mediterranean Sea and the Red Sea, and to some extent in the inner part of the Gulf of California, in which the bottom water has a high temperature and a high salinity and is formed by winter cooling of water whose salinity has been increased greatly by evaporation. Where such adjacent seas are in communication with the open oceans, deep water flows out over the sill, mixes with the water masses of the ocean, and spreads out at an intermediate depth corresponding to its density (p. 157).

In general, the *water of the greatest density is formed in high latitudes*, and, because this water sinks and fills all ocean basins, the deep and bottom water of all oceans is cold. Only in a few isolated basins in middle latitudes is relatively warm deep and bottom water encountered. When spreading out from the regions of formation the bottom water receives small amounts of heat from the interior of the earth, but this heat is carried away by eddy conduction and currents, and its effect on the temperature distribution is imperceptible.

Sinking of surface water is not limited to regions in which water of particularly high density is formed, but occurs also wherever converging currents (convergences) are present, the sinking water spreading at intermediate depths according to its density. In general, the density of the upper layers increases from the Tropics toward the poles, and water that sinks at a convergence in middle latitudes therefore spreads at a lesser depth than water that sinks at a convergence in high latitudes.

The most conspicuous convergence is the Antarctic Convergence, which can be traced all around the Antarctic continent (p. 155). The water that sinks at this convergence has a low salinity, but it also has a low temperature and consequently a relatively high density. This water, the Antarctic Intermediate Water, spreads directly over the deep water and is present in all southern oceans at depths between 1200 and 800 m. The corresponding Arctic Convergence is poorly developed in the North Atlantic Ocean, where an Atlantic Arctic Intermediate Water is practically lacking, but it is found in the North Pacific, where Pacific Arctic Intermediate Water is typically present.

In middle and lower latitudes, two more convergences are found—the Subtropical and the Tropical Convergences. These are not so well defined as the Antarctic Convergence, but must be considered more as *regions* in which converging currents are present. The Subtropical Convergence is located in latitudes in which the density of the upper layers increases rapidly toward the poles. The sinking water therefore has a higher density the farther it is removed from the Equator, and will spread out at the greatest depths.

In the Tropics the density of the surface water is so low that, regardless of how intense a convergence is, water from the surface cannot sink to any appreciable depth, but spreads out at a short distance below the surface. A sharp boundary surface develops between this light top layer and the denser water at some greater depth.

In order to account for the general features of the density distribution in the sea, emphasis has been placed on descending motion of surface water, but evidently regions must exist in which ascending motion prevails, because the amount of water that rises toward the surface must exactly equal the amount that sinks. Ascending motion occurs in regions of diverging currents (divergences), which may be present anywhere in the sea but which are particularly conspicuous along the western coasts of the continent, where prevailing winds carry the surface waters away from the coast. There the *upwelling* of subsurface water takes place, a process that will be described when dealing with specific areas. The upwelling brings water of greater density and lower temperature toward the surface, and therefore exercises a wide-spread influence upon conditions off coasts where the process takes place but where the water rises from depths of less than a few hundred meters. Ascending motion takes place on a large scale around the Antarctic Continent, particularly to the south of the Atlantic Ocean, where rising *deep water* replaces water that contributes to the formation of the Antarctic bottom water and of the water that sinks at the Antarctic Convergence.

It is evident from these considerations that in middle and lower latitudes the vertical distribution of density to some extent reflects the horizontal distribution at or near the surface between the Equator and the poles. It is also evident that, in general, the deeper water in any vertical column is composed of water from different source regions, and once was present in the surface layer somewhere in a higher latitude. Owing to the character of the currents, these generalizations are, however, subject to modifications within different ocean areas, and these modifications will be discussed when dealing with the different oceans.

Subsurface Distribution of Temperature and Salinity

The general distribution of temperature is closely related to that of the density. In high latitudes the temperature is low from the surface to the bottom. The bottom and deep waters that spread out from high latitudes retain their low temperature, but in middle and lower latitudes a warm top layer is present, the thickness of which depends partly on the processes of heating and cooling at the surface and partly on the character of the ocean currents. This upper layer of warm water is separated from the deep water by a layer of transition within which the temperature decreases rapidly with depth. From analogy with the atmosphere, Defant has applied the terms "troposphere" and "stratosphere" to two different parts of the ocean. "Troposphere" is applied to the upper layer of relatively high temperature which is found in middle and lower latitudes and within which strong currents are present, and "stratosphere" is applied to the nearly uniform masses of cold deep and bottom water. This distinction is often a very useful one, particularly when dealing with conditions in lower latitudes, but it must be borne in mind that the terms are based on an imperfect analogy with atmospheric conditions and that only some of the characteristics of the atmospheric troposphere and stratosphere find their counterparts in the sea.

So far, we have mainly considered an ideal ocean extending to high northerly and high southerly latitudes. Actually, conditions may be complicated by communication with large basins that contribute to the formation of deep water, such as the Mediterranean Sea, but these cases will be dealt with specifically when considering the different regions. Conditions will be modified in other directions in the Indian and Pacific Oceans, which are in direct communication with only one of the polar regions, and these modifications will also be taken up later. Here it must be emphasized that the general distribution of temperature is closely related to the distribution of density, which again is controlled by external factors influencing the surface density and the type of deep-sea circulation.

The general distribution of salinity is more complicated than that of temperature. Within the oceanic stratosphere the salinity is very uniform, but within the troposphere it varies greatly, being mainly related to the excess of evaporation over precipitation. The distribution of surface salinity, already discussed (p. 71), is in general characteristic of the distribution within the troposphere, as is evident from the vertical section in figs. 60 (p. 213) and 61 (p. 216).

Water Masses

THE *T-S* DIAGRAM. Water masses can be classified on the basis of their temperature-salinity characteristics, but density cannot be used for classification, because two water masses of different temperatures and salinities may have the same density. For the study of water masses, it is convenient to make use of the *temperature-salinity* (*T-S*) *diagram*, which was introduced by Helland-Hansen. Helland-Hansen pointed out that when in a given area the temperatures and corresponding salinities of the subsurface water are plotted against each other, the points generally fall on a well-defined curve, the *T-S* curve, showing the temperature-salinity relationship of the subsurface water of that region. Surface data have to be omitted, because annual variations and local modifications lead to discrepancies.

The corresponding temperature and salinity values in a water column are found to arrange themselves according to depth. The depths of the observed values can be entered along the *T-S* curve, which then will also give information as to variation of temperature and salinity with depth.

The density of the water at atmospheric pressure, which is expressed by means of σ_t (p. 11), depends only on temperature and salinity, and

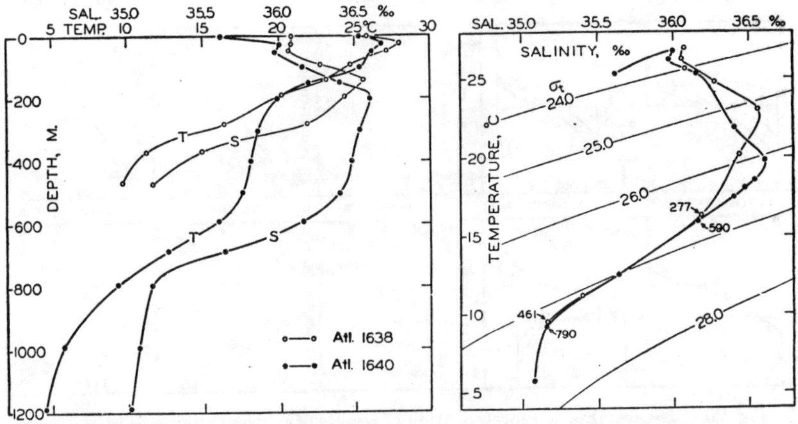

Fig. 19. *Left:* Temperature and salinity at *Atlantis* stations 1638 and 1640 in the Gulf Stream off Onslow Bay, plotted against depth. *Right:* The same data plotted in a *T-S* diagram in which σ_t curves have been entered.

therefore curves of equal values of σ_t can be plotted in the *T-S* diagram. If a sufficiently large scale is used, the exact σ_t value corresponding to any combination of temperature and salinity can be read off, and, if a small scale is used, as is commonly the case, approximate values can be obtained. The slope of the observed *T-S* curve relative to the σ_t curves gives immediately an idea of the stability of the stratification (p. 100).

A *T-S* diagram is shown on the right in fig. 19. On the left in the same figure the observed temperatures and salinities at *Atlantis* stations 1638 and 1640 in the Gulf Stream off Onslow Bay are plotted against depth, and on the right the same values are entered in a *T-S* diagram. The depths of a few of the observations are indicated. In this case, the temperature-salinity values between 277 and 461 m at station 1638 agree with those between 690 and 790 m at station 1640, indicating that at the two stations water of similar characteristics was present but at different depths.

The *T-S* diagram has become one of the most valuable tools in physical oceanography. By means of this diagram, characteristic features of the temperature-salinity distribution are conveniently represented and anomalies in the distribution are easily recognized. The diagram is also widely used for detecting possible errors in the determination of temperature or salinity (p. 44).

Water Masses and Their Formation. In accordance with Helland-Hansen's original suggestion a *water mass* is defined by a *T-S* curve, but in exceptional cases it may be defined by a single point on a *T-S* curve—that is, by means of a single temperature and a single salinity value. These exceptional cases are encountered in basins where homogeneous water is present over a wide range of depth. A *water type*,

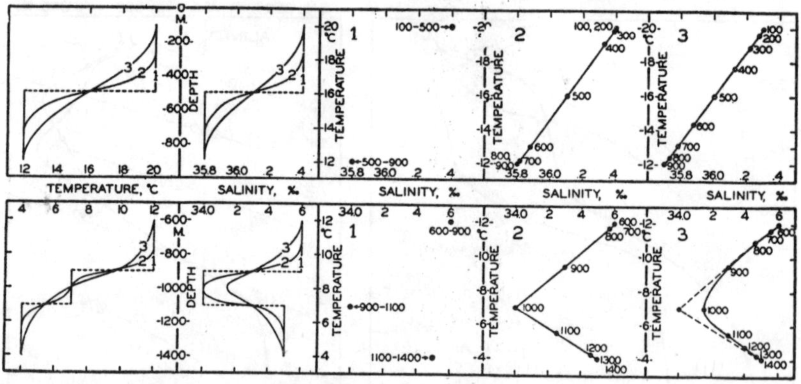

Fig. 20. Diagrammatic representation of results of vertical mixing of water types. To the left the results of mixing are shown by temperatures and salinities as functions of depth, and to the right are shown in three *T-S* diagrams the initial water types (1) and the *T-S* relations produced by progressive mixing (2 and 3).

on the other hand, is defined by means of single temperature and salinity values, but a given water type is generally present along a surface in the sea and has no thickness. Only in the exceptional cases that were referred to are the terms "water type" and "water mass" interchangeable, but in oceanographic literature the terms have been used loosely and without the distinction that has been introduced here.

In many areas the *T-S* curves are straight lines or can be considered as composed of several pieces of straight lines. Elementary considerations show that a linear relation between temperature and salinity must result if the water types, which can be defined by the end points of the straight line, mix in different proportions. Similarly, a curved *T-S* relation may result from the mixing of three different types of water. Fig. 20 illustrates in two simple cases how progressive mixing alters the temperature-salinity relation. These considerations are of a formalistic nature, but have in many instances led to the concept that certain water *types* exist and that the *T-S* relations that are observed represent the end results of mixing between the types. This concept presupposes that the water types (often referred to as water masses) are continually renewed, because, if that were not the case, processes of mixing would ultimately lead to the formation of homogeneous water. It is possible, however, to account for the character of the *T-S* curves in the ocean by considering other processes.

In the first place, it should be observed that a water mass of uniform temperature and salinity is rarely formed in the open ocean. In high latitudes, where convection currents in winter may reach to the bottom, the deep and bottom water will mostly not be uniform, because in some years the density of the surface water will be greater than in other years, and the convection current will reach to different depths, depending upon how much the density of the surface water has been increased. As a consequence, even in these areas the density increases toward the bottom; the bottom water is not homogeneous and shows, therefore, a definite temperature-salinity relationship. In the second place, sinking at convergences in middle latitudes may lead, as pointed out by Iselin, to the formation of a water mass with a *T-S* curve that reflects the horizontal distribution of temperature and salinity at the surface. Fig. 21A illustrates this point. The figure represents a schematic cross section in which are entered isotherms and isohalines that are all parallel and that all cut the surface. The σ_t curves have not been plotted, but are parallel to the isolines. The indicated system will remain stationary if sinking of surface water takes place between the lines *a* and *b* and if the sinking water remains on the same σ_t surface. It will also remain stationary if mixing takes place along or across σ_t surfaces. These processes will lead to the formation of a water mass which, between the curves *a* and *b*, always shows the same temperature-salinity relation—namely, the relation that is found along the sea surface. Iselin showed that the horizontal *T-S* curve along the middle part of the North Atlantic Ocean is very similar to the vertical *T-S* curve that is characteristic between temperatures of 20° and 8° over large areas of the North Atlantic Ocean, and he suggested that processes of sinking and lateral mixing are mainly responsible for the formation of that water. Extensive use of this concept will be made in the chapter dealing with the water masses and currents of the oceans.

However, a similar *T-S* relationship can be established by different processes, as is illustrated in fig. 21B. It is here assumed that two water types, *a* and *b*, are formed at the surface and sink, following their characteristic σ_t surfaces. It is furthermore assumed that, at subsurface depths, mixing takes place between these two water types, whereas, near the surface, external processes influence the distribution of temperature and salinity in such a way that the different curves cross each other. In these circumstances one obtains at subsurface depths a *T-S* relation that is similar to the one which was found in the previous example, but along the sea surface an entirely different *T-S* relation may exist. The processes of mixing must in this case take place across σ_t surfaces in order to establish the *T-S* relation, but at subsurface depths mixing along σ_t surfaces is not excluded. At present it is impossible to decide which process is of the greater importance.

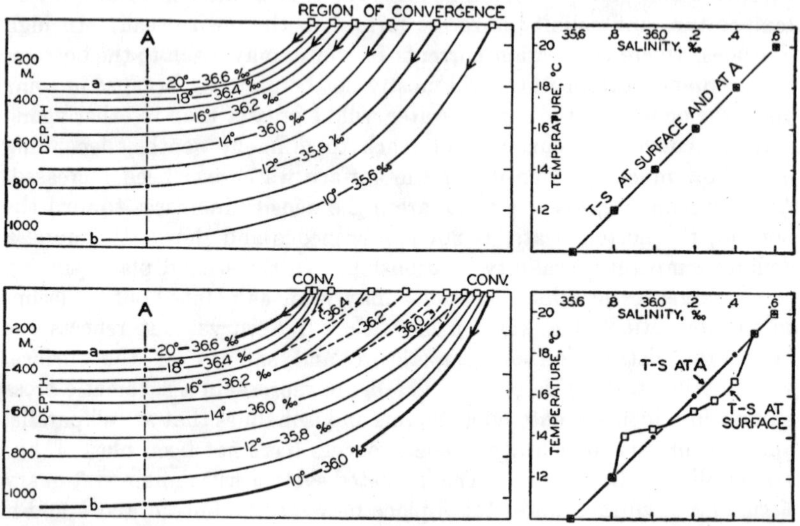

Fig. 21. A. Schematic representation of the formation of a water mass by sinking along σ_t surfaces (which coincide with the parallel temperature and salinity surfaces) in a region of convergence. The diagram to the right demonstrates that the vertical *T-S* relation of the water mass agrees with the horizontal *T-S* relation at the surface in the region of convergence. B. Schematic representation of the formation of a water mass by sinking of water types at two convergences and by subsequent sinking. The diagram to the right illustrates that in this case the vertical *T-S* relation of the water mass need not agree with the horizontal *T-S* relation at the surface between the convergences.

The point to bear in mind is that the waters of the ocean all attained their original characteristics when the water was in contact with the atmosphere or subject to heating by absorption of radiation in the surface layers, and that in course of time these characteristics may become greatly changed by mixing. This mixing can be either lateral—that is, take place along σ_t surfaces—or it can be vertical, crossing σ_t surfaces.

An example of lateral mixing between water masses is found off the coast of California, where the water that flows north close to the coast has a *T-S* relation which differs greatly from that of the water flowing south at some distance from the coast. Between these two water masses are found waters of an intermediate character that could not possibly have been formed by vertical mixing and must have been formed by lateral mixing, probably along σ_t surfaces. An example of modification of a water mass by vertical mixing is found in the South Atlantic where the Antarctic Intermediate Water flows north. This water, near its origin, is characterized by a low salinity minimum, but the greater the distance from the Antarctic Convergence the less pronounced is the salinity minimum (fig. 60, p. 213). This change probably cannot be accounted for by lateral mixing, but Defant has shown that it can be fully explained as a result of vertical mixing.

Wüst has introduced a different method for the study of the spreading out and mixing of water types—the "core method." By the "core" of a layer of water is understood that part of the layer within which temperature or salinity, or both, reach extreme values. Thus, in the Atlantic Ocean, the water that flows out from the Mediterranean has a very high salinity and can be traced over large portions of the Atlantic Ocean by means of an intermediate salinity maximum which decreases with increasing distance from the Strait of Gibraltar. The layer of salinity maximum is considered as the core of the layer in which the Mediterranean Water spreads, and the decrease of the salinity within the core is explained as the result of processes of mixing. In this case a certain water type, the Mediterranean Water, enters the Atlantic Ocean and loses its characteristic values, owing to the mixing, but can be traced over long distances. The spreading of the water can also be described by means of a *T-S* curve, one end point of which represents the temperature and salinity at the source region and the other end point the temperature and salinity in the region where the last trace of this particular water disappears. Having defined such a *T-S* curve, one can directly read off from the curve the percentage amount of the original water type that is found in any locality. The core method has proved very successful in the Atlantic Ocean, and is particularly applicable in cases in which a well-defined water type spreads out from a source region.

It has been mentioned that differences in density in a horizontal direction can exist only in the presence of currents, and therefore the relation between density distribution and currents must now be discussed. The character of the currents in the different oceans must also be examined, because knowledge of the oceanic circulation is necessary to an understanding of the interaction between the atmosphere and the sea.

CHAPTER VI

Ocean Currents Related to the Distribution of Mass

Introduction

The ocean currents may be conveniently divided into three groups: (1) currents that are related to the distribution of density in the sea, (2) currents that are caused directly by the stress which the wind exerts on the sea surface, and (3) tidal currents and currents associated with internal waves.

To the first class belong the well-known, large-scale ocean currents such as the Gulf Stream, the Kuroshio, the Equatorial Currents, the Benguela Current, and others. All of these currents transport great amounts of water. Their courses at the surface are known from ships' observations (p. 45), and at subsurface depths their character has, in a few localities, been determined from direct measurements of currents from anchored vessels, and in many more localities they have been ascertained from determinations of temperatures and salinities.

The wind drift also transports water in one and the same direction over large areas if it blows prevailing from one direction. The wind drift is of fundamental importance to the development and maintenance of the ocean circulation, and will therefore be dealt with in a separate chapter.

In contrast to these two types of currents the currents that are associated with tides and internal waves run alternatingly in opposite directions or are rotating. Although such currents may attain high velocities and are oceanographically significant, they are of no direct importance to the circulation of the ocean waters or to the interaction between the ocean and the atmosphere and will therefore not be discussed here.

In general, all three types of currents are present at the same time, and this makes it virtually impossible to obtain knowledge of the ocean currents on an entirely empirical basis. If such knowledge were to be obtained, it would be necessary to conduct measurements from anchored vessels at numerous localities for long periods and at many depths. By means of such series the different types of periodic currents could be examined, and by averaging they could be eliminated and the other types studied. In many oceanographic problems, however, knowledge of the periodic current is not essential, but knowledge of the currents that transport water over long distances is important. Currents related to the distribution of density can be computed from the more easily observed temperatures and salinities by following the procedure that will be discussed in detail in the following pages, and wind currents can be examined theoretically. Herein lie the value of the application of hydrodynamics to oceanography and the necessity of familiarity with this application if all possible conclusions are to be drawn from the observed distributions.

Equations of Motion Applied to the Ocean

General Equations. When dealing with ocean currents the same forces have to be considered as those which must be taken into account when discussing the dynamics of the atmosphere, namely (1) gravitational forces, (2) forces due to pressure gradients, (3) the deflecting force of the earth's rotation, and (4) frictional forces. In a *left-handed* rectangular coordinate system with the positive z-axis directed downward the equations of motion have the form

$$
\begin{aligned}
\frac{dv_x}{dt} &= \frac{\partial v_x}{\partial t} + v_x \frac{\partial v_x}{\partial x} + v_y \frac{\partial v_x}{\partial y} + v_z \frac{\partial v_x}{\partial z} = 2\Omega \sin \varphi \, v_y - \alpha \frac{\partial p}{\partial x} + \alpha R_x, \\
\frac{dv_y}{dt} &= \frac{\partial v_y}{\partial t} + v_x \frac{\partial v_y}{\partial x} + v_y \frac{\partial v_y}{\partial y} + v_z \frac{\partial v_y}{\partial z} = -2\Omega \sin \varphi \, v_x - \alpha \frac{\partial p}{\partial y} + \alpha R_y, \qquad \text{(VI, 1)} \\
\frac{dv_z}{dt} &= \frac{\partial v_z}{\partial t} + v_x \frac{\partial v_z}{\partial x} + v_y \frac{\partial v_z}{\partial y} + v_z \frac{\partial v_z}{\partial z} = -2\Omega \sin \varphi \, v_E + g - \alpha \frac{\partial p}{\partial z} + \alpha R_z.
\end{aligned}
$$

The symbols used have the meaning:

v_x, v_y, v_z,	the velocity components along the coordinate axes
v_E,	the horizontal velocity toward the east
Ω,	angular velocity of rotation of the earth, 0.729×10^{-4}
φ,	geographic latitude
α,	specific volume of the water
p,	pressure
R_x, R_y, R_z,	components of the frictional force per unit volume
g,	acceleration of gravity.

In the above form the equations are applicable to conditions in the Northern Hemisphere. For the Southern Hemisphere the sign of the first term on the right-hand side of the first two equations, the term representing the horizontal components of the deflecting force, must be reversed.

Motion in the Circle of Inertia. If the horizontal component of the deflecting force of the earth's rotation is the only acting force, the equations of horizontal motion have the form

$$\frac{dv_x}{dt} = 2\Omega \sin \varphi \, v_y,$$
$$\frac{dv_y}{dt} = -2\Omega \sin \varphi \, v_x. \qquad \text{(VI, 2)}$$

These equations define motion in a circle of radius r, where

$$r = \frac{v}{2\Omega \sin \varphi}, \qquad \text{(VI, 3)}$$

v being the horizontal velocity. The time of rotation in the inertia circle is $T = 2\pi/2\Omega \sin \varphi$, which is called "one half pendulum day." Motion of this type has been observed in the sea. The most striking example is found in a report by Gustafson and Kullenberg, in which are described the results of 162 hours' continuous record of currents in the Baltic. The measurements were undertaken between the coast of Sweden and the island of Gotland, in a locality where the depth to the bottom was a little over 100 m. On August 17, 1933, when the measurements began, a well-defined stratification of the water was found. From the surface to a depth of about 24 m the density increased rapidly with depth. Below 30 m a slow increase continued toward the bottom. The current meter, a Pettersson photographic recording meter, was suspended at a depth of 14 m below the surface, and thus would record the motion of the upper, homogeneous water.

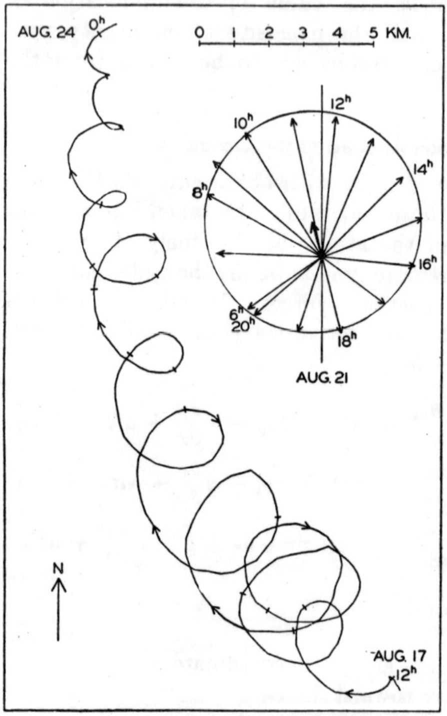

Fig. 22. Rotating currents of period one half pendulum day observed in the Baltic and represented by a progressive vector diagram for the period, August 17 to August 24, 1933, and by a central vector diagram between 6^h and 20^h on August 21 (according to Gustafson and Kullenberg).

Gustafson and Kullenberg have represented the results of the records in the form of a "progressive vector diagram" (fig. 22), which is prepared by successive graphical addition of the hourly displacements as computed from the average hourly velocities. Every twelfth hour is marked on the curve by a short line. The curve represents the path taken by a water particle, if it is assumed that the observed motion is characteristic of a considerably extended water mass.

The curve shows a general motion toward the northwest and later toward the north; superimposed upon this is a turning motion to the right, the amplitude of which first increases and later decreases. This rotation to the right (*cum sole*) is brought out by means of the inset central vector diagram in which the observed currents between August 21, 6^h and 20^h, are represented. The end points of the vectors fall nearly on a circle, as should be expected if the rotation is a phenomenon of inertia, but the center of the circle is displaced to the north-northwest, owing to the character of the main motion.

The period of one rotation was 14 hours, which corresponds closely to one half pendulum day, the length of which in the latitude of observation is 14^h08^m; on an average the periodic motion was nearly in a circle. It is possible that this superimposed motion can be ascribed to the effect of wind squalls, and that the gradual reduction of the radius of the circle of inertia is due either to frictional influence or to a spreading of the original disturbance. According to a theoretical examination by Defant, it is probable that inertia oscillations of this nature are associated with internal waves.

SIMPLIFIED EQUATIONS OF MOTION. When the general equations of motion are applied to the ocean, certain simplifications can be made. The vertical acceleration and the frictional term R_z can always be neglected. Similarly, the term depending upon the vertical component of the deflecting force can be neglected. The third equation of motion is therefore reduced to the *hydrostatic equation*, and the equations can be written in the following form, introducing the abbreviations $\lambda = 2\Omega \sin \varphi$, $\dot{v}_x = dv_x/dt$, and $\dot{v}_y = dv_y/dt$ and measuring the pressure in *decibars* (p. 10):

$$\begin{aligned} \dot{v}_x &= \lambda v_y - 10\alpha \frac{\partial p}{\partial x} + \alpha R_x, \\ \dot{v}_y &= -\lambda v_x - 10\alpha \frac{\partial p}{\partial y} + \alpha R_y, \\ 0 &= g - 10\alpha \frac{\partial p}{\partial z}. \end{aligned} \qquad \text{(VI, 4)}$$

At perfect hydrostatic equilibrium the isobaric surfaces coincide with level surfaces, but this is no longer the case if motion exists. At any time an isobaric surface is defined by

$$dp = \frac{\partial p}{\partial x}dx + \frac{\partial p}{\partial y}dy + \frac{\partial p}{\partial z}dz = 0. \qquad \text{(VI, 5)}$$

From equations (VI, 5) and (VI, 4) one obtains the equation for the isobaric surfaces in a moving system:

$$dp = (\lambda\rho v_y + R_x - \rho\dot{v}_x)dx + (-\lambda\rho v_x + R_y - \rho\dot{v}_y)dy + g\rho dz = 0. \qquad \text{(VI, 6)}$$

From equations (VI, 5) and (VI, 4) it also follows that the components of the horizontal pressure gradient are identical with the components of gravity acting along the isobaric surfaces:

$$g\left(\frac{dz}{dx}\right)_{p,x} = gi_{p,x} = \frac{-g\,\alpha\,\dfrac{\partial p}{\partial x}}{\alpha\,\dfrac{\partial p}{\partial z}} = -10\alpha\,\frac{\partial p}{\partial x},$$

$$g\left(\frac{dz}{dy}\right)_{p,y} = gi_{p,y} = \frac{-g\,\alpha\,\dfrac{\partial p}{\partial y}}{\alpha\,\dfrac{\partial p}{\partial z}} = -10\alpha\,\frac{\partial p}{\partial y}. \qquad \text{(VI, 7)}$$

The inclination is *positive* in the direction in which the surface slopes *downward* because the positive z axis is directed downward.

To the above equations must be added the equation of continuity and the kinematic and dynamic boundary conditions. The *equation of continuity* can be written in the form:

$$-\frac{1}{\rho}\frac{d\rho}{dt} = \frac{1}{\alpha}\frac{d\alpha}{dt} = \frac{\partial v_x}{\partial x} + \frac{\partial v_y}{\partial y} + \frac{\partial v_z}{\partial z} = \text{div } \boldsymbol{v}, \qquad \text{(VI, 8)}$$

and states that the rate of expansion per unit volume of a moving element equals the divergence of the velocity. The vector symbol $\boldsymbol{v}$ is used for the velocity. Water is nearly incompressible, and in the sea the vertical component of velocity is always small. When applied to the oceans the equation of continuity is often reduced to $(\partial v_x/\partial x + \partial v_y/\partial y) = 0$.

The *kinematic boundary condition* states that a particle on a boundary surface must move in a direction normal to the surface with the velocity of that surface. If the boundary surface is rigid, the water in contact with the surface can have no velocity component normal to the surface.

A *dynamic boundary condition* states that at any boundary surface the pressure must be the same on both sides of the surface. This also applies to internal boundaries that separate water of different density, in which case the condition states only that the pressure must vary continuously. The densities and velocities may, however, vary abruptly when passing from one side of the boundary surface to the other. Calling the densities on both sides ρ and ρ', and the velocities $\boldsymbol{v}$ and $\boldsymbol{v}'$, and omitting the frictional terms and the accelerations, the dynamic boundary condition takes the form

$$d(p - p') = \lambda(\rho v_y - \rho' v_y')dx - \lambda(\rho v_x - \rho' v_x')dy + g(\rho - \rho')dz = 0, \qquad \text{(VI, 9)}$$

because along the boundary surface $p = p'$, where p represents the pressure exerted against the boundary surface from one side, and p' represents the pressure exerted from the other side. Equation (VI, 9) must be fulfilled everywhere along the boundary surface, and defines therefore the shape of that surface in the same manner that (VI, 6) defines the isobaric surfaces.

The *dynamic energy equation* is obtained by considering that the work done by a force is equal to the product of the force and the distance traveled in the direction of the force. Multiplying the equations of horizontal motion by v_x and v_y, respectively, and adding, one obtains

$$\rho \frac{d}{dt}\left(\frac{1}{2}v^2\right) = -10\left(\frac{\partial p}{\partial x}v_x + \frac{\partial p}{\partial y}v_y\right) + R_x v_x + R_y v_y, \qquad \text{(VI, 10)}$$

because

$$v_x\dot{v}_x + v_y\dot{v}_y = \frac{d}{dt}\left(\frac{1}{2}v_x^2 + \frac{1}{2}v_y^2\right) = \frac{d}{dt}\left(\frac{1}{2}v^2\right),$$

and because the terms containing the deflecting force cancel. On the left-hand side of the equation stands the increase of kinetic energy per unit mass. On the right-hand side stands the sum of the work per unit mass performed by the forces due to pressure distribution and friction.

The equation is of small interest because it tells only that the increase of kinetic energy per unit mass equals the work performed per unit mass by the acting forces, but, combined with the *thermodynamic energy* equation, it becomes of importance. The complete derivation is given by V. Bjerknes and collaborators; only the result for a system that is enclosed by solid boundaries is stated here:

$$\frac{dW}{dt} = \frac{d}{dt}(K + \Phi + E) - \iiint (R_x v_x + R_y v_y)dx\,dy\,dz, \qquad \text{(VI, 11)}$$

where W is the amount of heat added to the system, K is the total kinetic energy of the system, Φ is the potential, and E is the internal energy. If the total energy of the system remains unaltered, the amount of heat added in unit time must equal the work per unit time of the frictional forces. If, on the other hand, no heat is added, the work of the frictional forces must lead to a change of the total energy of the system.

PRACTICAL APPLICATION OF THE HYDRODYNAMIC EQUATION FOR THE COMPUTATION OF OCEAN CURRENTS. In their above form the hydrodynamic equations are still too complicated to serve practical purposes, and hence they must be further simplified by assuming (1) that acceleration can be neglected and (2) that frictional forces can be neglected. On these assumptions the components of the horizontal velocities are obtained:

$$v_x = -\frac{10}{\lambda}\alpha\frac{\partial p}{\partial y} \quad \text{and} \quad v_y = \frac{10}{\lambda}\alpha\frac{\partial p}{\partial x}, \qquad \text{(VI, 12)}$$

or

$$v_x = \frac{1}{\lambda}g i_{p,y} \quad \text{and} \quad v_y = -\frac{1}{\lambda}g i_{p,x}. \qquad \text{(VI, 13)}$$

These equations are equivalent to the equations of the *geostrophic wind* in the atmosphere. The velocities are expressed here either by the pressure gradients, in which case the factor 10 enters because the pressure is measured in decibars, or by the inclination of the isobaric surfaces (VI, 7). It will be shown that the ocean currents can be obtained with sufficient accuracy from these simple equations, and the problem of computing the currents is reduced to a determination of the pressure gradients in the sea or of the inclination of the isobaric surfaces. Thus, if the field of pressure in the sea is known, the corresponding currents can also be approximately known.

The field of pressure can be fully represented by (1) a series of charts showing isobars in standard level surfaces or (2) a series of charts showing the topography of standard isobaric surfaces.

The term *level surface* is understood to mean a surface that is everywhere normal to the acceleration of gravity—that is, a surface of constant gravity potential. The work required for bringing a unit mass from one level surface to another is independent of the path taken. If the unit mass is moved the distance h along the plumb line, the work is $W = gh$, where g is the acceleration of gravity. The numerical value of this work depends upon the units used for length and time. If length is measured in meters and time in seconds, the unit of work per unit mass is a *dynamic decimeter*. Thus, differences in geopotentials are in the meter-ton-second system (m.t.s.) measured in dynamic decimeters, but the *dynamic meter* is the practical unit for measuring geopotential differences and is indicated by the symbol D. The difference in geopotential between the sea surface and a level surface at the geometrical depth, z, is therefore in dynamic meters:

$$D = \tfrac{1}{10} \int_0^z g\,dz, \qquad \text{(VI, 14)}$$

and similarly the geometric depth of a given level surface is

$$z = 10 \int_0^D \frac{1}{g}\,dD. \qquad \text{(VI, 15)}$$

Since the acceleration of gravity varies with latitude and depth, the *geometric* distance between level surfaces is variable, being greater at the Equator than at the poles and being greater near the sea surface than at great depths, but the "dynamic distance" between two given level surfaces is constant. When the pressure field is represented by the topography of isobaric surfaces, it is of advantage to use "dynamic" contours—that is, to represent the lines of intersection between the isobaric surfaces and a series of level surfaces. Charts of this character will be called charts of *geopotential topography*. If the dynamic meter is used as the unit of geopotential, the geopotential slope of an isobaric surface is

$$\frac{dD}{dx} = \frac{1}{10} g i_x, \qquad \text{(VI, 16)}$$

where i_x represents the geometric slope. Consequently the currents as computed from (VI, 13) are also represented by

$$v_x = \frac{10}{\lambda}\left(\frac{dD}{dy}\right) \quad \text{and} \quad v_y = -\frac{10}{\lambda}\left(\frac{dD}{dx}\right). \qquad \text{(VI, 17)}$$

That is, the current is parallel to the geopotential contours and is so directed that in the Northern Hemisphere, for which the above equations are valid, the surface rises to the right of an observer looking in the direction of flow, and in the Southern Hemisphere it rises to the left.

The Fields of Pressure and Mass in the Ocean

The relation between the distribution of mass and pressure is expressed by the hydrostatic equation, which can be written

$$dp = \rho_{s,\vartheta,p} dD \qquad \text{or} \qquad dD = \alpha_{s,\vartheta,p} dp, \qquad \text{(VI, 18)}$$

where the pressure is measured in decibars and the geopotential in dynamic meters. If the distribution of mass is known, the hydrostatic equation can be used for computing the *differences* in pressure at two dynamic depths, D_1 and D_2, or the *difference* in dynamic depth of two pressures, p_1 and p_2. In order to represent the pressure field completely, it is necessary to know the isobars in *one* level surface or the topography of *one* isobaric surface.

In dealing with the atmosphere, complete information can be obtained, because the distribution of pressure at sea level can be derived from direct observations of pressure and because the decrease of pressure with height can be computed from upper air observations of temperature and humidity. When dealing with the oceans, however, one encounters the fundamental difficulty that it is impossible to determine directly the pressure in any level surface or the topography of any isobaric surface. The sea surface itself can in the first approximation be taken as an isobaric surface, but the topography of the sea surface cannot be ascertained, although, according to oceanographic evidence, the sea surface in many localities is definitely sloping. Consider the fact that in the Gulf Stream off Cape Hatteras in latitude 35°N current velocities up to 150 cm/sec have been observed. From equations (VI, 13) it follows that the slope of the surface must be equal to 1.5×10^{-5}. The current flows toward the northwest, and the surface therefore rises toward the southeast, the rise being 1.5 cm in 1 km. This slope is computed from observations of the currents, and cannot be measured in any manner. Direct observations of the pressure along the sea bottom would be of value only if the points at which the pressure was observed could be joined by lines of precise leveling, but this is obviously impossible.

On the other hand, it is possible to determine the field of mass in the sea, since the density or the specific volume of the water can be obtained with great accuracy from observations of temperature and salinity. If the distribution of mass is known, the *relative* topography of the isobaric surfaces can be derived by using the hydrostatic equation, and the corresponding *relative* currents can be computed.

THE FIELD OF MASS. The field of mass in the ocean is generally described by means of the specific volume (p. 12):

$$\alpha_{s,\vartheta,p} = \alpha_{35,0,p} + \delta. \qquad \text{(VI, 19)}$$

The field of the specific volume can be considered as composed of two fields—the field of $\alpha_{35,0,p}$ and the field of δ. The former field is of a simple character. The surfaces of $\alpha_{35,0,p}$ coincide with the isobaric surfaces, the deviations of which from level surfaces are so small that for practical purposes the surfaces of $\alpha_{35,0,p}$ can be considered as coinciding with level surfaces or with surfaces of equal geometric depth. The field of $\alpha_{35,0,p}$ can therefore be fully described by means of tables giving $\alpha_{35,0,p}$ as a function of pressure and giving the average relationships between pressure, geopotential, and geometric depths. This field can be considered a constant one, and the field of mass is therefore completely described by means of the anomaly of the specific volume, δ, the determination of which was discussed on p. 12.

The field of mass can be represented by means of the topography of anomaly surfaces or by means of horizontal charts or vertical sections in which curves of δ = constant are entered. The latter method is the most common.

The density of the ocean waters generally increases with depth, and, therefore, the stratification is stable. The degree of stability, or, briefly, the *stability*, is evidently related to the change of density with depth and can be expressed as

$$E = \frac{1}{\rho}\frac{\delta\rho}{dz}, \qquad \text{(VI, 20)}$$

where $\delta\rho$ represents the difference in density between a small mass of water and its surroundings after the mass is moved adiabatically through the vertical distance dz. In the upper layers, $E = 10^{-3}d\sigma_t/dz$, approximately.

With regard to the field of mass, it must be added that equipotential surfaces do not exist in the sea—that is, no surfaces exist along which masses of water may be moved without altering the potential energy of the system. The σ_t surfaces are approximately equipotential surfaces, and in recent years considerable evidence has been accumulated to show that in the ocean lateral mixing takes place along σ_t surfaces and that the direction of flow is parallel to σ_t surfaces.

THE FIELD OF PRESSURE. If the field of mass is known, the relative field of pressure can be determined from the hydrostatic equation in one of the forms

$$dp = \rho dD \qquad \text{or} \qquad dD = \alpha dp.$$

In oceanography the latter form has been found to be the more practical, but all reasoning applies equally well to results deduced from the former.

Integration of the latter form gives

$$D_1 - D_2 = \int_{p_1}^{p_2} \alpha_{s,\vartheta,p} dp.$$

Because of (VI, 19), one can write

$$(D_1 - D_2)_s + \Delta D = \int_{p_1}^{p_2} \alpha_{35,0,p} dp + \int_{p_1}^{p_2} \delta dp, \qquad \text{(VI, 21)}$$

where

$$(D_1 - D_2)_s = \int_{p_1}^{p_2} \alpha_{35,0,p} dp \qquad \text{(VI, 22)}$$

is called the *standard* geopotential distance between the isobaric surfaces p_1 and p_2, and where

$$\Delta D = \int_{p_1}^{p_2} \delta dp \qquad \text{(VI, 23)}$$

is called the *anomaly* of the geopotential distance between the isobaric surfaces p_1 and p_2 or, abbreviated, the *geopotential anomaly.*

Equation (VI, 21) can be interpreted as stating that the relative field of pressure is composed of two fields—the standard field and the field of anomalies. The standard field can be determined once and for all, because the standard geopotential distance between isobaric surfaces represents the distance if the salinity of the sea water is constant at 35°/oo and if the temperature is constant at 0°C. It follows that a chart showing the topography of one isobaric surface relative to another by means of the geopotential *anomalies* is equivalent to a chart showing the actual geopotential topography of one isobaric surface relative to another. The practical determination of the relative field of pressure is therefore reduced to computation and representation of the geopotential anomalies, but the absolute pressure field can be found only if one can determine independently the absolute topography of one isobaric surface.

In order to evaluate equation (VI, 23) it is necessary to know the anomaly, δ, as a function of absolute pressure. The anomaly is computed from observations of temperature and salinity, but oceanographic observations give information about the temperature and the salinity at known geometrical depths below the actual sea surface, and not at known pressures. This difficulty can fortunately be overcome by means of an artificial substitution, because at any given depth the numerical value of the absolute pressure expressed in decibars is nearly the same as the numerical value of the depth expressed in meters, as is evident from the following corresponding values:

Standard sea pressure (decibars)...........	1000	2000	3000	4000	5000	6000
Approximate geometric depth (meters)......	990	1975	2956	3933	4906	5875

Thus, the numerical values of geometric depth deviate only 1 or 2 per cent from the numerical values of the standard pressure at that depth. This agreement is not accidental but has been brought about by the selection of the practical unit of pressure, the decibar.

TABLE 13

EXAMPLE OF COMPUTATION OF ANOMALIES OF DYNAMIC DEPTH
(Station *E. W. Scripps* I-8. Lat. 32°57′N, Long. 122°07′W. February 17, 1938)

Meters or decibars	Temp. (°C)	Salinity (°/₀₀)	σ_t	$10^5\Delta_{s,\vartheta}$	$10^5\delta_{s,p}$	$10^5\delta_{\vartheta,p}$	$10^5\delta$	ΔD	ΔD (dynamic meter)
0........	14.22	33.25	24.81	315.0			315		
10........	13.72	.24	.91	305.5		0.3	306	.0310	.0310
25........	.71	.24	.91	305.5		0.7	306	.0459	.0769
50........	.35	.30	25.03	294.2	−0.1	1.3	296	.0752	.1521
75........	9.96	.57	.86	215.2	−0.2	1.6	217	.0641	.2162
100........	.38	.84	26.17	185.7	−0.3	2.0	187	.0505	.2667
150........	8.82	.98	.37	166.5	−0.3	2.9	169	.0890	.3557
200........	.48	34.09	.51	153.5	−0.3	3.7	157	.0815	.4372
250........	.30	.16	.59	145.9	−0.4	4.6	150	.0768	.5140
300........	7.87	.20	.69	136.4	−0.5	5.2	141	.0728	.5868
400........	.07	.20	.80	125.9	−0.6	6.4	132	.1365	.7233
500........	6.14	.26	.97	109.8	−0.7	7.2	116	.1240	.8473
600........	5.51	.35	27.12	95.6	−0.8	7.9	103	.1095	.9568
800........	4.65	.42	.28	80.4	−1.0	8.9	88	.1910	1.1478
1000........	3.99	.44	.36	72.9	−1.2	9.8	82	.1700	1.3178
1200........	.52	.52	.48	61.5	−1.4	10.3	70	.1520	1.470
1400........	.07	.54	.54	55.8	−1.0	10.8	66	.1360	1.606
1600........	2.69	.56	.59	51.1	−1.0	10.9	61	.1270	1.733
1800........	.37	.59	.64	46.3	−1.0	10.8	56	.1170	1.850
2000........	.13	.64	.69	41.6	−1.1	10.8	51	.1070	1.957
3000........	1.62	.68	.76	35.0	−1.4	11.7	45	.4800	2.437
4000........	.50	.70	.81	30.2	−1.7	14.1	43	.4400	2.877

It follows that the temperature at a pressure of 1000 decibars is nearly equal to the temperature at a geometric depth of 990 m, or the temperature at the pressure of 6000 decibars is nearly equal to the temperature at a depth of 5875 m. The vertical temperature gradients in the ocean are small, especially at great depths, and therefore no serious error is introduced if, instead of using the temperature at 990 m when computing δ, one makes use of the temperature at 1000 m, and so on. The *difference* between anomalies for neighboring stations will be even less affected by this procedure, because within a limited area the vertical temperature gradients will be similar. The introduced error will be nearly the same at both stations, and the difference will be an error of absolutely negligible amount. In practice one can therefore consider *the numbers that represent the geometric depth in meters as representing absolute pressure in decibars.* Interpreting the depth in meters at which either directly observed or interpolated values of temperatures and salinities are available as representing *pressures* in decibars, one can compute, by means of a few simple tables, the anomaly of specific volume at the given pressure. By multiplying the average anomaly of specific volume between two pressures by the difference in pressure in decibars (which is considered equal to the difference in depth in meters), one obtains the geopotential anomaly of the isobaric sheet in question, expressed in dynamic meters. By adding these geopotential anomalies, one can find the corresponding anomaly between any two given pressures. An example of a complete computation is given in table 13.

As has been repeatedly stated, these computations give information only for the field of pressure that is related to differences in density. The total field of pressure may be composed of this internal field and of a field that may be related to piling up or removal of mass. If actual piling up of water takes place, the slopes of the isobaric surfaces remain constant from the surface to the bottom, and this pressure field is called the *slope field*, in contrast to the *internal field.*

Currents in Stratified Water

Some of the outstanding relationships between the distribution of mass and the velocity field are brought out by considering two water masses of different density, ρ and ρ', where ρ is greater than ρ'. In the absence of currents, the boundary surface between the two water masses is a level surface, the water mass of the lesser density, ρ', lying above the denser water. On the other hand, if the water of density ρ' moves at a uniform velocity v' and the water of greater density moves at another uniform velocity v, the boundary must slope. For the sake of simplicity, it will be assumed that both water masses move in the direction of the y axis and that the acceleration of gravity, g, can be considered constant. The dynamic boundary condition (p. 96) then takes the form

$$\lambda(\rho v - \rho' v')dx + g(\rho - \rho')dz = 0. \qquad \text{(VI, 24)}$$

From this equation one obtains the slope of the boundary surface

$$i_B = \frac{dz}{dx} = \frac{\lambda}{g}\frac{\rho' v' - \rho v}{\rho - \rho'}. \qquad \text{(VI, 25)}$$

The slope of the isobaric surfaces is obtained from (VI, 6), which, on the above assumptions, gives

$$i'_p = -\frac{\lambda}{g}v' \quad \text{and} \quad i_p = -\frac{\lambda}{g}v. \qquad \text{(VI, 26)}$$

The slope of the isobaric surfaces is small compared to the slope of the boundary surface, but even the latter is very small under conditions encountered in the ocean.

As an extreme case, consider two water masses, one of salinity 34.00°/₀₀ and temperature 20°, and one of salinity 35.00°/₀₀ and temperature 10°, and assume that the velocity of the former is 0.5 m/sec. Neglecting the effect of pressure on density, one obtains:

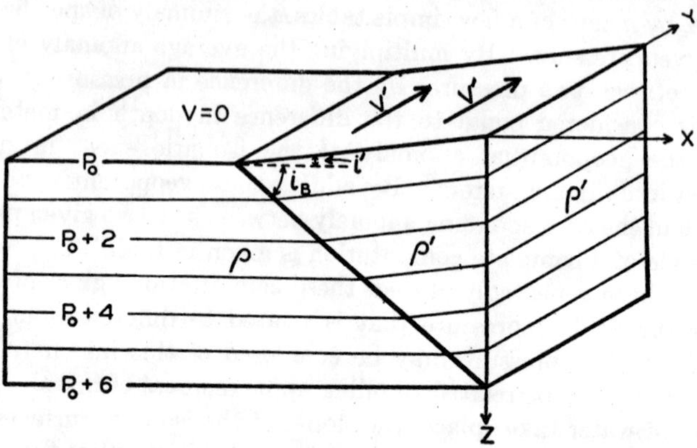

Fig. 23. Isobaric surfaces and currents within a wedge of water that extends over resting water of greater density.

$$\rho' = 1.02402, \qquad \rho = 1.02697, \qquad v' = 50 \text{ cm/sec}, \qquad v = 0.$$

In latitude 40°N, where $\lambda = 0.937 \times 10^{-4}$ sec^{-1} and $g = 980$ cm/sec^2, formula (VI, 25) gives

$$i_B = 1.66 \times 10^{-3};$$

that is, the boundary surface sinks 1.66 m when x, the horizontal distance, increases by 1 km (the positive z axis is directed downward).

The slopes of the isobaric surfaces are much smaller. Inserting the numerical values in (VI, 26), one obtains

$$i'_p = -0.47 \times 10^{-5} \qquad \text{and} \qquad i_p = 0,$$

meaning that within the upper layer the isobaric surfaces rise 0.47 m on a horizontal distance of 100 km and that within the lower layer they are horizontal. The conditions are represented schematically in the block diagram in fig. 23, where the slope of the isobaric surfaces is greatly exaggerated. Actually, the lighter water extends like a thin wedge over the heavier water.

As another example, consider the case where the current in the upper layer is limited to a band of width L. In this case, perfect static equilibrium must exist in the regions of no currents, and there the boundary between the two water masses must be horizontal, but, in the region where the water of low density flows with a velocity v', the boundary surface must slope, the steepness of the slope being determined by (VI, 25). These conditions are shown schematically by the block diagram in fig. 24, where it is supposed that the denser water reaches the surface on the left-hand side of the current. This figure brings out an important relation-

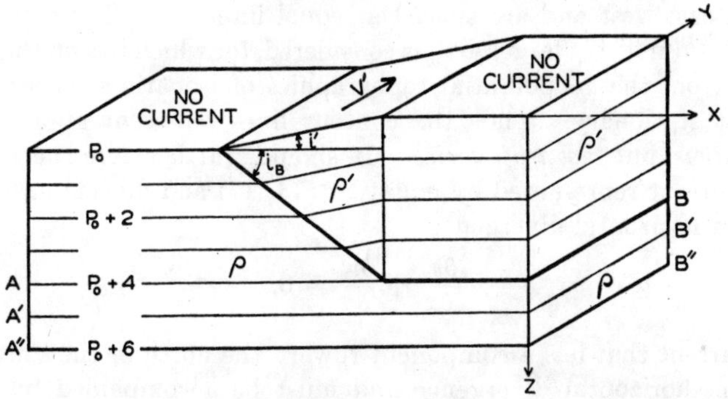

Fig. 24. Isobaric surfaces and currents within water that in part extends as a wedge over resting water of greater density.

ship between the current and the distribution of mass: *The current flows in such a direction that the water of low density is on the right-hand side of the current and the water of high density is on the left-hand side.* This rule applies to conditions in the Northern Hemisphere, since in the Southern Hemisphere the water of low density will be to the left and the water of high density will be to the right.

Practical Methods for Computing Ocean Currents

"RELATIVE" CURRENTS. As has already been stated (p. 99), oceanographic observations can give information only as to the relative topographies, and therefore only as to the corresponding "relative" velocities. When using the term "relative" velocity, it should be borne in mind that the actual velocities (relative to the earth) are not obtained by adding a constant value to the "relative," but that the value that has to be added in order to obtain the actual velocities varies from one vertical to another, depending upon the unknown slope of the isobaric surface that has been used as a reference surface.

The contour lines of the isobaric surfaces represent the stream lines of the relative motion, but they are not, as a rule, identical with trajectories, even if the reference surface is level so that the computed velocities represent the actual motion. If such is the case, the contour lines will be trajectories only if the motion is stationary—that is, if $\partial v_x/\partial t = \partial v_y/\partial t = 0$ (p. 93). Because it has been assumed that $\dot{v}_x = \dot{v}_y = 0$, it follows that the conditions

$$v_x \frac{\partial v_x}{\partial x} + v_y \frac{\partial v_x}{\partial y} = 0,$$

$$v_x \frac{\partial v_y}{\partial x} + v_y \frac{\partial v_y}{\partial y} = 0$$

must be fulfilled. Examination of these conditions shows that the motion can be stationary only if the geopotential contour lines are directed east-west and are spaced at equal intervals. These conditions are never fulfilled if a larger area is considered, for which reason the motion derived from the geopotential topographies of isobaric surfaces is, as a rule, not stationary. Thus the contour lines represent approximately *stream lines* but *not trajectories.* It should, furthermore, be observed that a current represented by equation (VI, 17) and flowing east-west is free from horizontal divergence:

$$\frac{\partial v_x}{\partial x} + \frac{\partial v_y}{\partial y} = 0,$$

but a current that has a component toward the north or the south is *not* free from horizontal divergence and must be accompanied by vertical motion.

The computation of the geopotential distances between isobaric surfaces has already been discussed, and the whole problem of computing ocean currents would therefore be very simple if (1) simultaneous observations of temperature and salinity at different depths were available from a number of stations so that relative topographies could be constructed, (2) accelerations could be neglected, (3) frictional forces could be neglected, and (4) periodic changes in the distribution of mass as related to internal waves were negligible.

Simultaneous observations from a number of stations are never available, and the question therefore arises as to whether charts based on stations that have been occupied within a certain time interval can be considered as approximately representing a synoptic situation. This question can be examined by repeated surveys of the same area. Such surveys have shown that conditions vary in time, but so slowly that the main features of a certain topography are represented correctly by nonsimultaneous observations that have been taken within a reasonably short time interval. Results of repeated surveys are found in the publications of the U. S. Coast Guard presenting the work of the International Ice Patrol off the Grand Banks of Newfoundland, where a small area has been covered in less than a week and where cruises have been repeated at intervals of three to four weeks. In these intervals of time the details of the relative topography have changed greatly, but the main features have changed much more slowly. Similar surveys have been conducted by the Scripps Institution of Oceanography off southern California.

Fig. 25 shows the results of two cruises off the coast of southern California in 1940. The upper chart shows the topography of the surface relative to the 500-decibar surface according to observations between

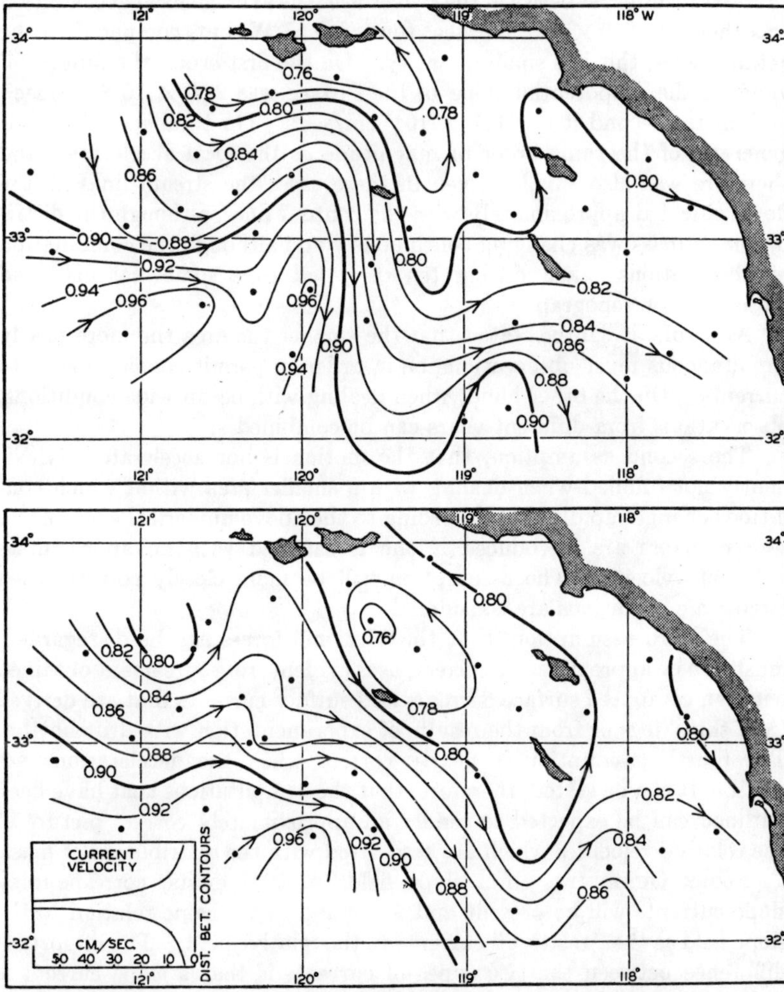

Fig. 25. Geopotential topography of the sea surface in dynamic meters referred to the 500-decibar surface, according to observations off southern California on April 4 to 14, 1940 (upper), and on April 22 to May 3, 1940 (lower).

April 4 and 14, and the lower chart shows the corresponding topography on April 22 to May 3. The stations upon which the charts are based are shown as dots. The time interval between the cruises was about sixteen days, and, although the main features were similar, the details were greatly changed. Thus, in latitude 33°40′N, longitude 120°W, the computed surface current on the first cruise was 36 cm/sec toward N 80°E, whereas on the second cruise it was 14 cm/sec toward S 30°E. In the time interval between the two cruises the average local acceleration was therefore 2.7×10^{-5} cm/sec^2 toward S 57°W, but, compared to the acting forces, this is a small quantity. On the first cruise the numerical value of the geopotential slope of the surface was 294×10^{-5} cm/sec^2, and on the second it was 117×10^{-5} cm/sec^2. The total acceleration is generally of the same order of magnitude as the local acceleration and therefore was also small. Even in these cases the stream lines of the flow coincided approximately with the contour lines, although the distribution of mass was changing continuously and although the combination of observations taken during ten days led to a somewhat distorted picture of the topography.

As a rule, it can be stated that the smaller the area the more nearly simultaneous must observations be in order to permit conclusions as to currents. On the other hand, when dealing with ocean-wide conditions, observations from different years can be combined.

The second assumption, that the motion is not accelerated, is evidently not fulfilled when dealing with a smaller area within which conditions change rapidly, but according to the above numerical example no serious errors are introduced if one is satisfied with an approximate value of velocity. The assumption will be more closely correct when large-scale conditions are considered.

The third assumption, that the frictional forces can be disregarded, must also be approximately correct, as is evident from agreement obtained between computed surface currents and surface currents that are derived from ships' logs or from the results of experimentation with drift bottles. The fourth assumption is nearly correct when the accelerations are small. It can be stated, therefore, that the computations that have been outlined can be expected to render an approximately correct picture of the relative velocities which are associated with the distribution of mass.

SLOPE CURRENTS. If a slope field (p. 103) exists, corresponding slope currents will be present and will stand in the same relation to the slope field as the "relative" currents to the relative field. The important difference between the two types of currents is that a slope current is uniform from the sea surface to the bottom, whereas a relative current changes with depth.

Existence of slope fields and corresponding slope currents has been demonstrated in such land-locked bodies of water as the Baltic Sea, and results of precise leveling along the North American east coast suggest that slope currents may be present there also. In the open oceans, slope currents probably do not exist, except as transitory phenomena caused by changing winds (p. 133).

Actual Currents. So far, we have considered mainly the "relative" currents associated with the relative field of pressure, but the ultimate goal must be to determine the actual currents. The problem of determining actual currents can be dealt with in two steps. In the first place, one can consider whether there are reasons to assume that the actual currents are determined completely by differences in density. If this question is answered in the affirmative, one can then consider what reference surface should be used in order to find the actual motion.

The first question can be approached in the following manner: If the distribution of mass remains stationary, the flow must always be parallel to the isopycnals, because, if this condition is not fulfilled, the distribution of mass will be altered by the motion. On the assumptions made, the flow is always parallel to the isobars, and it follows, therefore, that under stationary conditions isopycnals and isobars must be parallel at all levels. It also follows that the isobars and isopycnals at one level must be parallel to those at all other levels. This rule is identical with the "law of the parallel solenoids" of Helland-Hansen and Ekman. The motion that has to be considered when dealing with distortion of the field of mass depends, however, on the total field of pressure, and this total field must evidently have the same geometrical shape as the internal field if the law of the parallel solenoid shall be fulfilled. The total field is composed of the internal and the slope fields (p. 103), and consequently these fields must coincide if the law of the parallel solenoids shall be fulfilled. It is very unlikely, however, that a slope field of such character develops, for which reason parallel isopycnals and isobars strongly indicate that a slope field is absent.

The study of large-scale conditions in the ocean has shown that over large areas the isotherms, isohalines, and, consequently, the isopycnals are parallel at different levels and that their direction coincides with the direction of the relative isobars or the contour lines of the isobaric surfaces. This empirical result strongly supports the view that the large-scale currents are largely determined by the internal distribution of mass. Even in small areas a similar arrangement is often found; but many exceptions are encountered there which clearly demonstrate that stationary conditions do not exist, and which may be interpreted as indicating that the currents are *not* determined entirely by the distribution of mass.

The next question that arises is whether it is possible in the ocean to determine a surface along which the velocity is zero, so that actual velocities are found when the "relative" motion is referred to this surface. Such a surface need not be an isobaric surface but may have any shape.

One school of oceanographers points out that the deep waters of the oceans are nearly uniform and that the isopycnal surfaces there are nearly horizontal. It is therefore assumed that in the deep water the isobaric surfaces are also nearly horizontal and that actual currents are found if the reference level is placed at a sufficiently great depth. This method is useful if a study of the water masses substantiates the view that the motion of the deep water is negligible.

A second school of oceanographers claims that the distribution of oxygen in the ocean must be closely related to the type of motion, and especially that the layer or layers of minimum oxygen content that are found over large areas must represent layers of minimum horizontal motion, but this concept is finding a more and more limited application.

A third method has been employed by Defant, who points out that in the Atlantic Ocean the relative distances between isobaric surfaces remain nearly constant within certain intervals of depth. He assumes that a surface of no motion lies within this interval, and arrives in this manner at a consistent picture of the shape of the reference surface in the Atlantic and at results which are in good agreement with those obtained by considering the equation of continuity.

A fourth method is based on the equation of continuity, but this method has so far been little used because it requires comprehensive data. The application can be illustrated by considering the currents of such an ocean as the South Atlantic. It is evident that the net transport of water (p. 115) through any cross section of the South Atlantic between South Africa and South America must be zero, because water cannot permanently be removed from the North Atlantic or be accumulated in that ocean, which is practically a closed bay. In the South Atlantic a surface of no motion must therefore be selected in which the flow to the north above that surface equals the flow to the south below the same surface (p. 116). Similarly the surface of no motion has to be selected in other regions in such a manner that one arrives at a consistent picture of the currents, taking into account the continuity of the system and the fact that subsurface water masses retain their character over long distances. Examples of such pictures are shown in figs. 42 and 56.

In many regions the surface of no motion does not coincide with a level surface, and in these cases the topographies of isobaric surfaces have to be constructed stepwise by considering that the isobaric surfaces must be horizontal where they cross the surface of no motion.

From the many reservations made, it may appear that the computations reach only rough approximations to the actual conditions. However, in many instances in which comparisons are possible it has been found that the computed currents do not deviate much from the observed ones. As an example, in fig. 26 a comparison is shown between the velocities of the current through the Straits of Florida according to Wüst's computations and according to Pillsbury's detailed observations in the 1880's. The observed distribution of temperature and salinity and the corresponding currents are shown in parts *A*, *B*, and *C*, and in part *D* are shown the observed currents. The agreement is striking.

An example of the practical application of the methods is presented by the work of the International Ice Patrol, which is conducted by the U. S. Coast Guard. Oceanographic work off the Grand Banks of Newfoundland is carried out by a vessel of the U. S. Coast Guard in March and April, and on the basis of the observed temperatures and salinities between the surface and a depth of 1000 m the geopotential topography

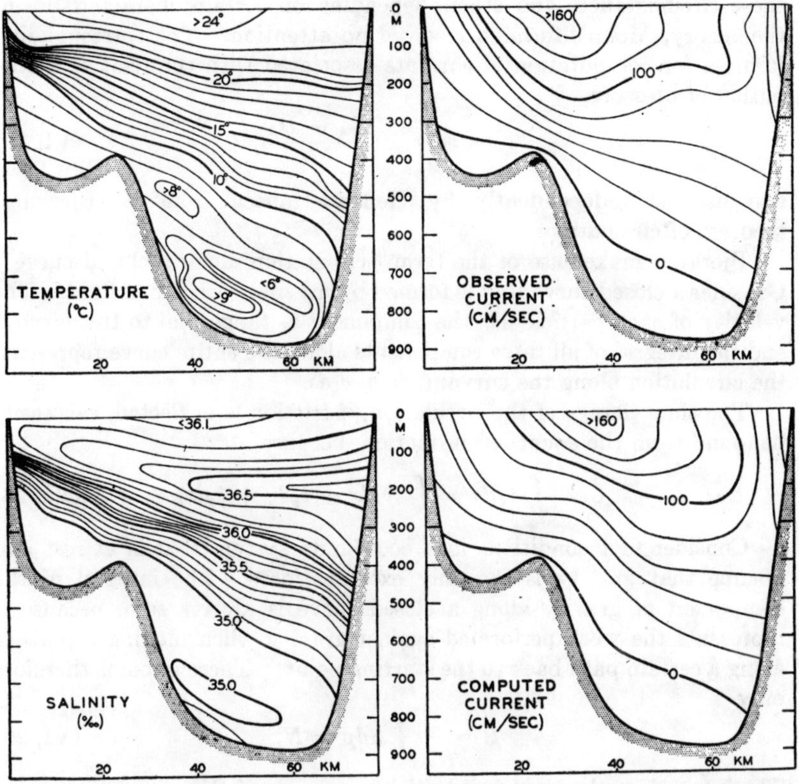

Fig. 26. *Left:* Observed temperatures and salinities in the Straits of Florida. *Right:* Velocities of the current through the Straits according to direct measurements and according to computations based on the distributions of temperature and salinity (after Wüst).

of the surface relative to the 1000-decibar surface is computed. Assuming that the flow follows the contour lines, the direction and speed of drifting icebergs are computed and warnings are issued to shipping. The method has proved sufficiently successful to warrant the extensive work year after year. In other regions in which computed currents have been compared to surface currents that have been derived from ship's observations or from the results of experiments with drift bottles, satisfactory agreement has been obtained. Thus, in spite of deficiencies of the method the computation of currents is a most useful tool in oceanography.

Bjerknes' Theorem of Circulation. The general formula for computing ocean currents from the slope of the isobaric surfaces, $v = -gi/\lambda$, was derived by H. Mohn in 1885, but at the time when Mohn presented his theory the oceanographic observations were not sufficiently accurate for computation of the relative field of pressure. Owing to these circumstances and others depending on certain characteristics of the theory, Mohn's formula received no attention. The corresponding formula for computation of currents associated with the relative distribution of pressure,

$$v_1 - v_2 = 10\,\frac{D_A - D_B}{\lambda L}, \qquad \text{(VI, 27)}$$

was derived independently by Helland-Hansen from V. Bjerknes' theorem of circulation.

Bjerknes makes use of the term "circulation along a closed curve." Consider a closed curve that is formed by moving particles of fluid. The velocity of each particle has the component v_T tangential to the curve c, and the integral of all these components along the entire curve represents the circulation along the curve: $C = \int_c v_T ds$.

The time change of the circulation, if friction is neglected, can easily be found from the equations of motion, because

$$\dot{C} = \int_c \dot{v}_T ds = \int_c (\dot{v}_x dx + \dot{v}_y dy + \dot{v}_z dz). \qquad \text{(VI, 28)}$$

Consider first conditions in a coordinate system that is at rest and assume that gravity is the only external force. The integral of the component of gravity along a closed curve is always zero, because it represents the work performed against gravity when moving a particle along a certain path back to the starting point. There remains therefore only

$$\dot{C} = -\int_c \alpha dp = N. \qquad \text{(VI, 29)}$$

The integral on the right-hand side is zero only if the specific volume is constant along the curve or is a function of pressure only. In these cases no internal field of force exists, and the theorem states that the circulation along a closed curve is constant if the fluid is homogeneous or if isosteric surfaces coincide with isobaric surfaces. If the isosteric surfaces cut the isobaric surfaces, the space can be considered as filled by tubes, the walls of which are formed by isosteric and isobaric surfaces. If these are entered with unit difference in specific volume and pressure, respectively, the tubes are called *solenoids*. It can be shown that the integral in equation (VI, 29) is equal to the number of solenoids, N, enclosed by the curve. Consider now a curve that runs vertically down to the isobaric surface $p = p_2$, and finally along this surface back to the starting point (fig. 27). Along the isobaric surfaces, $dp = 0$, and therefore

$$N = -\int_c \alpha\, dp = \left[\left(-\int_{p_1}^{p_2} \alpha\, dp\right)_B + \left(-\int_{p_2}^{p_1} \alpha\, dp\right)_A\right]$$
$$= 10\,(D_A - D_B). \quad \text{(VI, 30)}$$

If circulation relative to the earth is considered, the deflecting force has to be taken into account. According to Bjerknes, this is represented by

$$\dot{C}_r = -\int_c \alpha\, dp - 2\,\Omega\dot{\Sigma}, \quad \text{(VI, 31)}$$

where Σ represents the projection on the equatorial plane of the area that is enclosed by the curve c, and Ω, as before, means the angular velocity of rotation of the earth.

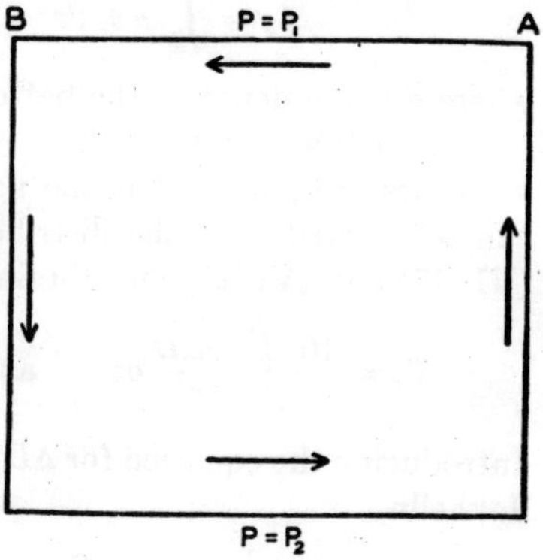

Fig. 27. Location in the pressure field of a curve along which the time change of circulation is examined.

Consider the same curve as before, and assume that the upper line, $p = p_1$, moves at a velocity v_1 at right angles to the line A-B, whereas the lower line, $p = p_2$, moves at a velocity v_2, and let the distance A-B be called L. Then the time change of the projection on the equator plane is

$$\dot{\Sigma} = \frac{v_1 - v_2}{L \sin \varphi}.$$

Assume now that the circulation is constant ($\dot{C} = 0$). It then follows that the velocity difference $v_1 - v_2$ must be expressed by the equation

$$v_1 - v_2 = 10\,\frac{D_A - D_B}{\lambda L}. \quad \text{(VI, 32)}$$

This is the equation for computing "relative" currents which Helland-Hansen derived from the theorem of circulation, assuming stationary conditions and neglecting friction, and which can be arrived at on the basis of more elementary consideration (VI, 17).*

The complete theorem of circulation contains a great deal more than the simple statement expressed by equation (VI, 32), but so far it has not been possible to make greater use in oceanography of Bjerknes'

* In the literature, some confusion exists as to the signs in these equations. The above signs are consistent with the coordinates used. The "relative" velocity is positive—that is, the "relative" current is directed away from the reader if A lies to the right of B and if D_A is greater than D_B.

elegant formulation of one of the fundamental laws governing the motion of nonhomogeneous fluids.

TRANSPORT BY CURRENTS. The *volume transport* by horizontal currents has the components

$$T_x = \int_0^d v_x \, dz \quad \text{and} \quad T_y = \int_0^d v_y \, dz, \tag{VI, 33}$$

and the *mass transport* has the components

$$M_x = \int_0^d \rho \, v_x \, dz \quad \text{and} \quad M_y = \int_0^d \rho \, v_y \, dz, \tag{VI, 34}$$

where d is the depth to the bottom.

The actual volume transport can be computed only if the actual velocities are known, but the volume transport by "relative" currents can be derived from the distribution of mass. By means of equations (VI, 17) and (VI, 23) one obtains

$$T_x = \frac{10}{\lambda} \int_0^d \frac{\partial \Delta D}{\partial y} dz \quad \text{and} \quad T_y = -\frac{10}{\lambda} \int_0^d \frac{\partial \Delta D}{\partial x} dz. \tag{VI, 35}$$

Introducing the equation for ΔD (VI, 23) and writing, in accordance with Jakhelln,

$$Q = \int_0^p \int_0^d \delta \, dp \, dz,$$

one obtains

$$T_x = \frac{10}{\lambda} \frac{\partial Q}{\partial y} \quad \text{and} \quad T_y = -\frac{10}{\lambda} \frac{\partial Q}{\partial x}. \tag{VI, 36}$$

The quantity Q is easily computed, because ΔD is always determined if velocities are to be represented.

In practice the numbers that represent the geometric depths are also considered as representing the pressures in decibars (p. 102). When computing Q the integration is therefore carried to the pressure p decibars and the depth d meters, which are both expressed by the same number, but a small systematic error is thereby introduced. Jakhelln has examined this error and has shown that the customary procedure leads to Q values that are systematically 1 per cent too small, but this error is negligible.

Curves of equal values of Q can be drawn on a chart, and these curves will bear the same relation to the "relative" volume transport that the curves of ΔD bear to the "relative" velocity, provided that the derivations of Q are taken as positive in the direction of decrease. Therefore the direction of the volume transport above the depth to which the Q values are referred will be parallel to the Q lines, and numerical values will be proportional to the gradient of the Q lines. The factor of proportionality will depend upon the latitude, however, and, between two parallel Q lines that indicate transport to the north, the transport will decrease in the direction of flow, whereas, between two parallel Q lines indicating transport to the south, the transport will decrease in the direction of flow. Taking this fact into account, one can construct curves between which the transport is constant, but these curves will no longer be exactly parallel to the direction of the transport. Within a small area the deviation will be imperceptible, but within a larger area it will be considerable.

Between two stations A and B (A lying to the right) at distance L the volume transport is

$$T = L\int_0^d v dz = \frac{10}{\lambda}\int_0^d (\Delta D_A - \Delta D_B)dz. \qquad \text{(VI, 37)}$$

Thus, the volume transport between two stations depends only on the geopotential anomalies at the two stations. It is independent of the distance between the stations, and it is also independent of the distribution of mass in the interval between the stations.

It is not necessary to develop corresponding equations for the *mass* transport, because this can be derived with sufficient accuracy from the volume transport by multiplication with an average density.

A consideration of computed transport and of the continuity of the system is helpful under certain conditions in separating currents flowing one above the other. The average transport through a vertical section that represents the opening to a basin must be zero, because water cannot accumulate in the basin, nor can it continue to flow out of the basin. Observations at stations at the two sides of such a section may, however, show transport in or out if the velocity along the bottom is assumed to be zero. If such is the case, the assumption is wrong, and a depth of no motion must be determined in such a manner that the transports above and below that depth are equal. This can be done by plotting the difference $\Delta D_A - \Delta D_B$ against depth, computing the average value of the difference between the surface and the bottom, and reading from the curve the depth at which that average value is found. This depth is then

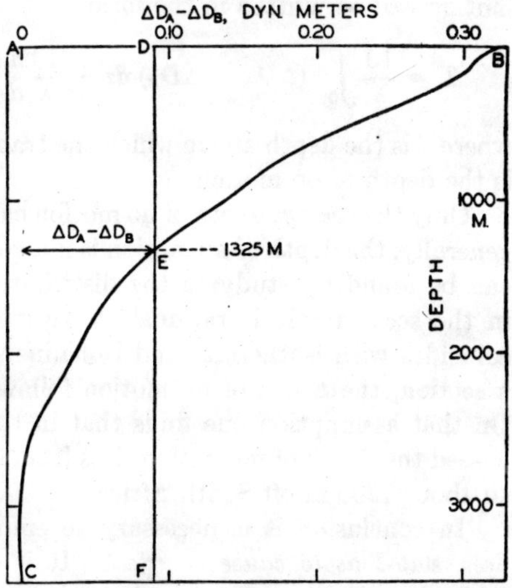

Fig. 28. Determination of the depth above and below which the transport between two stations, A and B, is equal.

the average depth of no motion, and the transports above and below that depth are in opposite directions.

The procedure is illustrated in fig. 28, in which is plotted the difference $\Delta D_A - \Delta D_B$ at *Meteor* station 36 off South America and station 23 off South Africa, both in about latitude 29°S. Assuming no current at a depth of 3500 m, one would obtain a transport to the north proportional to the area *ABC*, or equal to 44.5 million m³/sec. Such a transport to the north is impossible, and the velocity must therefore be zero at some other depth. The average value of $\Delta D_A - \Delta D_B$ is 0.090 dynamic meters, and this value is found at a depth of 1325 m. If this depth is selected as a depth of no motion, the transport to the north becomes proportional to the area *DBE* and is equal to 20.7 million m³/sec; below 1325 m the transport is directed to the south, is proportional to the area *ECF*, and is also equal to 20.7 million m³/sec. The transport above the depth of no motion can be written in the form

$$T = \frac{10}{\lambda}\int_0^h (\Delta D_A - \Delta D_B)\,dz - \frac{10}{\lambda}\frac{h}{d}\int_0^d (\Delta D_A - \Delta D_B)\,dz, \qquad \text{(VI, 38)}$$

where d is the depth above which the transport was first calculated and h is the depth of no motion.

Only the *average* depth of no motion can be determined in this manner; generally, the depth of no motion is not constant, but a variation in depth can be found by studying the distribution of temperature and salinity in the section. It is rational to assume that a surface of no motion coincides with isothermal and isohaline surfaces, and that therefore, in a section, the depth of no motion follows the isotherms and isohalines. On that assumption one finds that in the example which has been discussed the depth of no motion rises from about 1450 m off South America to about 1200 m off South Africa.

In conclusion it is necessary to emphasize that, so far, *nothing has been stated as to cause or effect*. It is obvious that all the equations which have been used state only that, on the assumptions made, a definite relationship exists between the currents and the distribution of mass. In order to discuss the *cause* of the ocean currents, it is necessary to examine the manner in which the distribution of mass is maintained. It is found that the distribution of mass is controlled by processes of radiation at the sea surface, the exchange of heat with the atmosphere, the evaporation from the sea surface, and the character of the prevailing winds. The processes of heating and cooling are all important in establishing and maintaining differences in density, but the wind renders the energy that is needed for maintaining the circulation. In contrast to the atmosphere the ocean is an inefficient thermodynamic machine that converts very small amounts of heat to kinetic energy.

CHAPTER VII

Wind Currents and Wind Waves

Frictional Forces

In the discussion of the currents related to the distribution of mass, frictional forces were disregarded, but they must be taken into account when considering the effect of the wind.

Two of the fundamental concepts concerning friction in a fluid are (1) that shearing *stresses* are produced when layers are slipping relative to each other, and (2) that the shearing stresses acting on a unit area are proportional to the rate of shear normal to the surface on which the stress is exerted. Thus the horizontal stress parallel to the x axis is $\tau_x = \mu dv_x/dz$, where the factor of proportionality, μ, is the dynamic viscosity of the fluid. The quantity $\nu = \mu/\rho$ is the kinematic viscosity.

Frictional *forces* per unit volume are equal to the difference between shearing stresses exerted on opposite sides of a cube of unit dimensions. Introducing differentials, one obtains the x component of the frictional force per unit volume:

$$R_x = \frac{d\tau_x}{dz} = \mu \frac{d^2v_x}{dz^2}, \qquad \text{(VII, 1)}$$

if μ is a constant.

In classical hydrodynamics the dynamic viscosity, μ, is considered to be a characteristic property of the fluid—namely, the property that resists angular deformation. Being a characteristic property, the magnitude of μ is independent of the state of motion, but as a rule it varies with the temperature of the fluid. Within wide limits it is independent of the pressure.

In a fluid in *turbulent* motion, much larger shearing stresses develop (Reynold's stresses) which are related to transport of momentum that is caused by irregular exchange of mass between neighboring layers moving with different mean velocities. If the horizontal components of the mean velocities are called $\bar{v}_x$ and $\bar{v}_y$, the components of the Reynold's stresses are

$$\tau_x = \mu_e \frac{d\bar{v}_x}{dz} \quad \text{and} \quad \tau_y = \mu_e \frac{d\bar{v}_y}{dz}.$$

The coefficient μ_e has the same dimensions as the dynamic viscosity μ, and is called the *eddy viscosity*. However, a fundamental difference exists between the two quantities. The dynamic viscosity is independent of the state of motion and is a characteristic property of the fluid, comparable to the elasticity of a solid body, but the eddy viscosity depends upon the state of motion and is not a characteristic physical property of the fluid. The numerical value of the eddy viscosity varies within very wide limits, according to the type of motion, and, as far as ocean currents are concerned, only the order of magnitude of μ_e has been ascertained (table 1, p. 23).

In the theory of turbulence the concept of the *mixing length* has played a prominent part. This length can be defined as the average distance which the small masses travel before they attain the momentum of their surroundings. According to Prandtl's theory the relation between mixing length, l, and eddy viscosity can be written

$$\mu_e = \rho l^2 \left|\frac{d\bar{v}}{dz}\right|. \tag{VII, 2}$$

According to von Kármán's general statistical theory,

$$\sqrt{\frac{\tau}{\rho}} = k_0 \frac{\left(\frac{d\bar{v}}{dz}\right)^2}{\frac{d^2\bar{v}}{dz^2}}, \tag{VII, 3}$$

where k_0 is a nondimensional universal constant that has been found equal to nearly 0.4.

So far, only horizontal shearing stresses have been considered, but recently evidence has been accumulated showing that in the ocean vertical shearing stresses also exist. It has been found that coefficients of horizontal mixing must be introduced which are so great that the corresponding stresses cannot be neglected (table 3, p. 25).

In dealing with wind currents, only the terms related to vertical mixing have been considered, but it is probable that a complete theory of the dynamics of ocean currents cannot be developed without taking horizontal mixing into account.

The Stress of the Wind

Methods used in fluid mechanics for studying frictional stresses at solid boundary surfaces can be applied, as was first shown by Rossby, in examining the stress that the wind exerts on the sea surface. Over an absolutely smooth surface the flow will generally be laminar within a very thin layer near the surface—the *laminar boundary* layer. Above this layer, the thickness of which is a small fraction of a centimeter, turbulent motion exists. On the assumption that near the boundary surface the stress, τ_0, is practically independent of the distance from the surface, von Kármán showed that the mixing length, l, increases linearly with increasing distance from the surface: $l = k_0 z$, where $k_0 = 0.4$, and that the relation between the velocity distribution in the turbulent region and the stress can be written in the form

$$\frac{W_z}{\sqrt{\frac{\tau_0}{\rho}}} = 5.5 + 5.75 \log \frac{z\rho}{\mu} \sqrt{\frac{\tau_0}{\rho}}, \qquad \text{(VII, 4)}$$

where μ is the dynamic viscosity. Applied to the sea surface, the theorem shows that, if a laminar boundary layer exists, the stress in the lowest layers, which must equal the stress against the sea surface, could be derived from a single measurement of the wind velocity, W, at a short distance, z, from the sea surface. Measurements at two or more levels would have to render the same value of the stress, and the observed velocities would have to be a linear function of the logarithm of the distance.

Over a *rough* surface, different conditions are encountered. Prandtl assumes that then the turbulent motion extends to the very surface, meaning that the mixing length has a definite value at the surface itself:

$$l = k_0(z + z_0), \qquad \text{(VII, 5)}$$

where z_0 is called the *roughness length* and is related to the average height of the roughness elements. According to Prandtl the eddy viscosity, μ_e, is expressed by (VII, 2). Therefore,

$$\tau_0 = \mu_e \frac{dW}{dz} = \rho k_0^2 (z + z_0)^2 \left(\frac{dW}{dz}\right)^2,$$

or

$$\frac{dW}{dz} = \frac{1}{k_0(z + z_0)} \sqrt{\frac{\tau_0}{\rho}} \qquad \text{and} \qquad \mu_e = \rho k_0 (z + z_0) \sqrt{\frac{\tau_0}{\rho}}.$$

Through integration, assuming $k_0 = 0.4$ and $W = 0$ at $z = 0$, and introducing base-10 logarithms, one obtains

$$W_z = 5.75 \sqrt{\frac{\tau_0}{\rho}} \log \frac{z + z_0}{z_0} \qquad \text{(VII, 6)}$$

and

$$\tau_0 = \rho \frac{0.0302}{\left(\log \frac{z + z_0}{z_0}\right)^2} W_z^2. \qquad \text{(VII, 7)}$$

Measurements of the wind velocity of two distances are needed in order to determine both τ_0 and z_0, and measurements at three or more levels are necessary in order to test the validity of the equation.

From wind measurements at different distances above the sea surface conducted by Wüst, Rossby and Montgomery concluded that at moderate wind velocities the sea surface has the character of a rough surface with roughness length 0.6 cm. From a general theory of the wind profile in the lower part of the atmosphere, Rossby later developed formulas that permit computation of z_0 from the angle between the surface wind and the gradient wind, or from the ratio between surface wind velocity and gradient wind velocity. When applying these formulas to observed values from the ocean, Rossby found that at moderate and strong winds the roughness length is independent of the wind velocity and is equal to 0.6 cm. With this value of z_0, one obtains from formula (VII, 7)

$$\tau_0 = 2.6 \times 10^{-3} \rho_a W_{15}^2, \qquad \text{(VII, 8)}$$

where ρ_a is the density of the air and W_{15} is the wind velocity 15 m above the sea surface.

As early as 1905, Ekman derived a similar expression by an entirely different approach, which will be explained on p. 122. Ekman's conclusions have been confirmed by the recent studies of Palmén and Laurila, but all results apply to the stresses exerted by moderate or strong winds.

At low wind velocities, somewhat different conditions appear to prevail. From his analyses, Rossby concludes that at low wind velocities the sea surface has the character of a smooth surface, because the wind profiles indicate the existence of a laminar sublayer. At low wind velocities the stress would therefore be computed from equation (VII, 4), leading to values which are about one third of the values obtained on the assumption that the surface is rough.

TABLE 14

STRESS OF THE WIND (G/CM/SEC²) CORRESPONDING TO STATED WIND VELOCITIES (IN M/SEC) AT A HEIGHT OF 15 M AND ASSUMING THE SURFACE TO BE SMOOTH OR ROUGH (ROUGHNESS LENGTH, 0.6 CM)

Surface	Wind velocity in m/sec at 15 m								
	2	4	6	8	10	12	14	16	18
"Smooth"...	0.04	0.16	0.34	(0.58)					
"Rough"....	(0.11)	(0.45)	(1.01)	1.81	2.83	4.09	5.56	7.25	9.20

Table 14 shows the values of the stress corresponding to a smooth surface or a rough surface characterized by $z_0 = 0.6$ cm. It is evident that, if the sea surface can be considered smooth at wind velocities below 6 or 7 m/sec, this limit is of little significance, but the above conclusions need to be confirmed by many more data. The problem of the stress of the wind still deserves great attention; it is particularly desirable to obtain more measurements of wind profiles, because results in fluid mechanics can be successfully applied in the study of such profiles.

The theoretical equations for the relations between stress and wind velocity are valid only if the stability of the air is nearly indifferent. Under stable or unstable conditions the wind profile and the relation between wind and stress will be altered. A study of wind profiles at different stabilities of the air is therefore of great interest.

Piling Up of Water Due to the Stress of the Wind

At the sea surface the stress that the wind exerts on the water, τ_a, must balance the stress that the water exerts on the air. The latter has the components $\mu_{e,0}(dv_x/dz)_0$ and $\mu_{e,0}(dv_y/dz)_0$, where μ_e now means the eddy viscosity of the water. Therefore,

$$\tau_{a,x} = -\mu_{e,0}\left(\frac{dv_x}{dz}\right)_0 \quad \text{and} \quad \tau_{a,y} = -\mu_{e,0}\left(\frac{dv_y}{dz}\right)_0. \tag{VII, 9}$$

This relation is useful in a study of the effect of wind based on the equations of motion of the water. Including the terms that determine friction due to vertical turbulence and assuming that acceleration can be neglected, the equations of motion take the form

$$\begin{aligned} \lambda\rho v_y + g\rho i_{p,x} + \frac{d}{dz}\left(\mu_e \frac{dv_x}{dz}\right) &= 0, \\ -\lambda\rho v_x + g\rho i_{p,y} + \frac{d}{dz}\left(\mu_e \frac{dv_y}{dz}\right) &= 0. \end{aligned} \tag{VII, 10}$$

Here the geometric slopes of the isobaric surfaces are introduced. Integrating the equations between the surface and the bottom and considering that the stress, τ_d, which the current exerts on the bottom has the components

$$\tau_{d,x} = \mu_{e,d}(dv_x/dz) \quad \text{and} \quad \tau_{d,y} = \mu_{e,d}(dv_y/dz), \tag{VII, 11}$$

one obtains

$$\begin{aligned} \tau_{a,x} + \tau_{d,x} &= -\lambda \int_0^d \rho v_y dz - \int_0^d g\rho i_{p,x} dz, \\ \tau_{a,y} + \tau_{d,y} &= \lambda \int_0^d \rho v_x dz - \int_0^d g\rho i_{p,y} dz. \end{aligned} \tag{VII, 12}$$

Consider next a channel which is so narrow that transverse motion can be neglected, and place the x axis in the direction of the channel ($v_y = 0$). Assume furthermore that the water is homogeneous, so that the inclination of the isobaric surface is independent of depth. The first of the above equations is then reduced to

$$\tau_{a,x} + \tau_{d,x} = -g\rho i_{p,x} d; \tag{VII, 13}$$

that is, the stresses acting in the direction of the channel at the upper and lower surfaces of the body of water are balanced by the component of gravity acting on the entire body of water (fig. 29).

This is the relation that Ekman used for determining the stress of the wind. Studying sea levels recorded during a storm in 1872 at a number of stations along the shores of the Baltic Sea, Colding found that the lines of equal sea level were nearly perpendicular to the wind direction, and that the relation between wind velocity, W, depth, d, and slope of the sea surface, i, could be expressed by the equation

$$id = 4.8 \times 10^{-9} W^2,$$

where depth is in centimeters and wind velocity in centimeters per second. Ekman showed that the value of τ_d probably lies between 0 and $\frac{1}{3}\,\tau_a$. Assuming $\tau_d = \frac{1}{3}\,\tau_a$, he obtained

$$\tau_a = 3.2 \times 10^{-6} W^2,$$

or, introducing the density of the air, $\rho_a = 1.25 \times 10^{-3}$,

$$\tau_a = 2.6 \times 10^{-3} \rho_a W^2.$$

From observations during a storm in the Gulf of Bothnia in 1936, Palmén and Laurila obtained similarly

$$\tau_a = 2.4 \times 10^{-3} \rho_a W^2,$$

and they consider this formula valid at wind velocities between 10 and 26 m/sec. These results are in very good agreement with Rossby's formula (VII, 8).

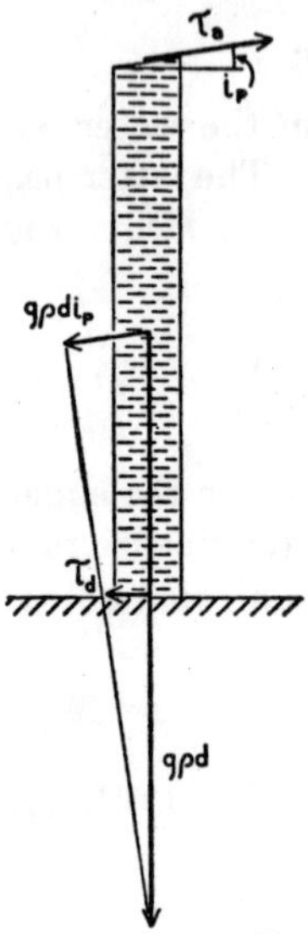

Fig. 29. Schematic representation showing the stresses acting at the upper and lower surfaces of a body of water balanced against the component of gravity acting on the entire water body.

Changes in sea level due to the direct effect of the stress of the wind are common in such shallow and enclosed areas as the Baltic Sea and the Gulf of Bothnia, and are often observed in shallow lakes. On open coasts where shallow waters extend to appreciable distances, similar piling up of water may take place during violent storms, causing wide inundation of low-lying areas. For instance, this took place during the hurricane of Galveston in 1900 and the hurricane of New England in 1938.

The above equations are not applicable to conditions in the open ocean, because there the density varies with depth and therefore the slopes of the isobaric surfaces vary. When dealing with the open ocean, it is more convenient to introduce the geopotential slope of the isobaric surfaces and write (VII, 13) in the form

$$\tau_{a,x} = -10 \int_0^d \rho \frac{d\Delta D}{dx}\, dz. \qquad \text{(VII, 14)}$$

The stress against the bottom, τ_d, has been neglected, because in the open ocean the velocities at great depths are negligible. In the ocean, slopes produced by the wind can be recognized, because the wind leads to a piling up of the lighter surface water. Thus, Montgomery and Palmén have shown that the trade winds of the North Atlantic and the North Pacific lead to a piling up of the warmer and lighter water masses against the western boundaries of the oceans. In the Atlantic Ocean the effect reaches to a depth of 150 m, and in the Pacific to a depth of 300 m. Below these depths the isobaric surfaces are horizontal, but above they are inclined, sloping from west to east, so that the trade winds actually blow uphill, but the slopes are small and decrease rapidly with depth. In the Atlantic the slope of the sea surface is 3.8×10^{-8}, and in the Pacific it is 4.5×10^{-8}. It appears that only part of the wind stresses are balanced against the slope, so that the complete equation (VII, 12) must be considered if the total effect of the wind is to be discussed.

In the calm belt between the trade-wind regions, no stress is exerted on the surface. Here the water must flow downhill as a countercurrent flowing toward the east between the west-flowing currents of the trade-wind regions. These equatorial countercurrents are completely reflected in the distribution of mass. Where they are located in the Northern Hemisphere the lighter water lies on the right-hand side and the denser water on the left-hand side of the current. This is a striking example of a case in which the distribution of mass is maintained by the current, and not *vice versa*.

Wind Currents in Homogeneous Water

In homogeneous water where no piling up of water takes place, equation (VII, 12) is reduced to

$$M_x = \frac{1}{\lambda}\tau_{a,y} \quad \text{and} \quad M_y = -\frac{1}{\lambda}\tau_{a,x}, \tag{VII, 15}$$

provided that no stress is exerted against the bottom. Here M_x and M_y represent the components of the mass transport between the surface and the bottom (p. 114), and the equations state that, regardless of the character of the viscosity, the total transport due to the stress of the wind is directed at right angles to the stress, in the Northern Hemisphere to the right, in the Southern Hemisphere to the left.

In order to examine the character of the wind currents, it is necessary to return to the equations of motion, which, on the above assumption, are reduced to

$$\begin{aligned} \lambda\rho v_y + \frac{d}{dz}\left(\mu_e \frac{dv_x}{dz}\right) &= 0, \\ -\lambda\rho v_x + \frac{d}{dz}\left(\mu_e \frac{dv_y}{dz}\right) &= 0. \end{aligned} \tag{VII, 16}$$

These equations were first integrated by Ekman, on the assumption that the eddy viscosity is a constant. Ekman undertook the mathematical analysis on the suggestion of Fridtjof Nansen, who, during the drift of the *Fram* across the Polar Sea in 1893–1896, had observed that the ice drift deviated 20 to 40 degrees to the right of the wind, and had attributed this deviation to the effect of the earth's rotation. Nansen further reasoned that, in a similar manner, the direction of the motion of each water layer must deviate to the right of the direction of movement of the overlying water layer, because it is swept on by this layer much as the ice which covers the surface is swept on by the wind; therefore, at some depth the current would run in a direction opposite to the surface flow. Nansen's conclusions were fully confirmed by mathematical treatment.

Assuming that the eddy viscosity is independent of depth, the equations can be directly integrated. If

$$D = \pi\sqrt{\frac{2\mu_e}{\lambda\rho}}, \tag{VII, 17}$$

the result is

$$\begin{aligned} v_x &= C_1 e^{\frac{\pi}{D}z}\cos\left(\frac{\pi}{D}z + c_1\right) + C_2 e^{-\frac{\pi}{D}z}\cos\left(\frac{\pi}{D}z + c_2\right), \\ v_y &= C_1 e^{\frac{\pi}{D}z}\sin\left(\frac{\pi}{D}z + c_1\right) - C_2 e^{-\frac{\pi}{D}z}\sin\left(\frac{\pi}{D}z + c_2\right), \end{aligned} \tag{VII, 18}$$

where C_1, C_2, c_1, and c_2 are constants that must be determined by means of the boundary conditions.

In this form the result is applicable to conditions in the Northern Hemisphere only, because in the Southern Hemisphere $\sin\varphi$ is negative, wherefore D is imaginary. In order to obtain a solution that is valid in the Southern Hemisphere, the direction of the positive y axis must be reversed.

The solution takes the simplest form if the depth to the bottom is so great that one can assume no motion near the bottom, because then C_1 must be zero. Assuming, furthermore, that the stress of the wind, τ_a, is directed along the y axis, one has

$$-\mu_e\left(\frac{dv_y}{dz}\right)_0 = \tau_a \quad \text{and} \quad \mu_e\left(\frac{dv_x}{dz}\right)_0 = 0,$$

from which equations C_2 and c_2 can be determined. Calling the velocity at the surface v_0, one obtains

$$\begin{aligned} v_x &= v_0 e^{-\frac{\pi}{D}z}\cos\left(45° - \frac{\pi}{D}z\right), \\ v_y &= v_0 e^{-\frac{\pi}{D}z}\sin\left(45° - \frac{\pi}{D}z\right), \\ v_0 &= \frac{\tau_a}{\sqrt{\mu_e\lambda\rho}} = \frac{2\pi\tau_a}{D\lambda\rho\sqrt{2}}. \end{aligned} \tag{VII, 19}$$

Therefore, the wind current is directed 45 degrees *cum sole* from the direction of the wind. The angle of deflection increases regularly with depth, so that at the depth $z = D$ the current is directed opposite to the surface current. The velocity decreases regularly with increasing depth, and at $z = D$ is equal to $e^{-\pi}$ times the surface velocity, or one twenty-third of the value at the surface. By far the more important velocities occur above the depth $z = D$, and Ekman has therefore called this depth "the depth of frictional resistance." It should be observed, however, that according to this solution the velocity of the wind current never becomes zero, but approaches zero asymptotically, being practically zero below $z = D$.

A schematic representation of the pure wind current is given in fig. 30. The broad arrows represent the velocities at depths of equal intervals. Together they form a spiral staircase, the steps of which rapidly decrease in width as they proceed downward. Projected on a horizontal plane, the end points of the vectors lie on a logarithmic spiral.

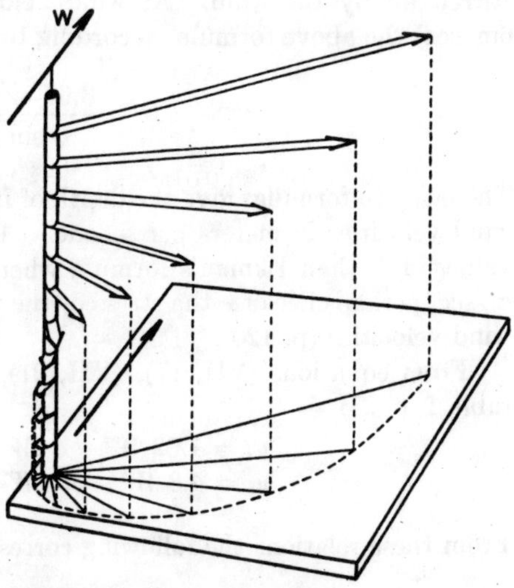

Fig. 30. Schematic representation of a wind current in deep water, showing the decrease in velocity and change of direction at regular intervals of depth (the Ekman spiral). W indicates direction of wind.

The average deflection of the wind current from the direction of the wind has been examined (Krümmel) and found to be about 45° *cum sole* from the wind direction, independent of latitude, in agreement with Ekman's theory.

The ratio between the velocity of the wind and that of the surface current, which Thorade has called the *wind factor*, depends upon the stress of the wind and upon the value of the eddy viscosity—that is, upon D. The stress of the wind is proportional to the square of the wind velocity, W_1; that is, $\tau_a = 3.2 \times 10^{-6} W^2$ (p. 120), where W is measured in centimeters per second. Introducing this value into (VII, 19), one obtains, with $\lambda = 2\Omega \sin \varphi$,

$$\frac{v_0}{W} = \frac{\pi\, 3.2 \times 10^{-6} W}{D \rho \Omega \sin \varphi \sqrt{2}}.$$

On the basis of observations by Mohn and Nansen, Ekman derived the empirical relation

$$\frac{v_0}{W} = \frac{0.0127}{\sqrt{\sin\varphi}}, \qquad \text{(VII, 20)}$$

from which follows

$$D = 7.6\,\frac{W}{\sqrt{\sin\varphi}}, \qquad \text{(VII, 21)}$$

and this value is in fair agreement with observed values of the upper homogeneous layer in the sea that is interpreted as the layer which is stirred up by the wind. At wind velocities below 3 Beaufort (about 6m/sec) the above formula, according to Thorade, should be replaced by

$$D = \frac{3.67\,\sqrt{W^3}}{\sqrt{\sin\varphi}}. \qquad \text{(VII, 22)}$$

The last two formulas give the depth of frictional resistance in meters for wind velocities in meters per second. The latter formula gives smaller values of D than Ekman's formula when the wind velocity is below 4.3 m/sec, perhaps because the stress of the wind is relatively smaller at low wind velocities (p. 120).

From equations (VII, 17), (VII, 21), and (VII, 22), one obtains (see table 1, p. 23)

$$\mu_e = 1.02\,W^3 \qquad (W \leq 6 \text{ m/sec}),$$
$$\mu_e = 4.3\,W^2 \qquad (W > 6 \text{ m/sec}).$$

From these relations the following corresponding values are computed:

Wind velocity (m/sec)	2	4	6	8	10	15	20
μ_e (g cm/sec)	8	65	(218)	375	430	970	1720

Rossby and Montgomery have developed a new theory of the wind currents by introducing an eddy viscosity depending upon a mixing length that is small near the surface, increases to a maximum at a short distance below the surface, and then decreases linearly to zero at the lower boundary of the wind-stirred layer. The variation with depth of the mixing length is assumed to be entirely determined by means of dimensionless numerical constants. The theory leads to a series of complicated expressions for the angle between surface current and stress, the wind factor, and the depth of the wind current. These quantities all depend both on latitude and on wind velocity. Rossby and Montgomery give their final results in the form of tables, according to which the deflection of the wind current in latitude 5° increases from 35 degrees at a wind velocity of 5 m/sec to 43 degrees at a wind of 20 m/sec, and in latitude 60° from 42 to 52.7 degrees. In the same latitude the wind factors decrease from 0.0317 at a wind velocity of 5 m/sec to 0.0266 at a velocity of 20 m/sec and from 0.0273 to 0.0228, respectively.

According to Ekman's theory the angle should remain constant at 45 degrees, and the wind factor, according to the empirical results that he used, should be equal to 0.025 in latitude 15° and 0.0136 in latitude 60°. Rossby and Montgomery point out that the wind factor depends upon the depth at which the wind current is measured. Their theoretical values apply to the current at the very surface, but wind currents derived from ships' logs will apply to a depth of 2 or 3 m, depending upon the draft of the vessel. At this depth, their theory gives a wind factor in better agreement with Ekman's value, and they show that their theoretical conclusions are in fair agreement with empirical results. However, the introduction of an eddy viscosity that decreases to zero at the lower limit of the wind-stirred layer is justifiable only if no other currents are present. In the presence of other currents, such as tidal currents or currents related to distribution of mass, the eddy viscosity characteristic of the total motion must be introduced, and the variation of this eddy viscosity probably depends more upon the stability of the stratification than upon the geometric distance from the free surface. A theory of the wind currents must take this fact into account, and cannot be developed until more is known of the actual character of the turbulence. At present, Ekman's classical theory appears to give a satisfactory approximation, especially because no observations are as yet available by means of which the results of a refined theory can be tested.

So far, it has been assumed that the depth of the water is great compared to the depth of frictional resistance. Ekman has also examined the wind currents in shallow water and has determined the constants in (VII, 18) by assuming that at the bottom the velocity is zero. This analysis leads to the result that in shallow water the deflection of the surface current is less than 45 degrees and the turning with depth is slower. In very shallow water the current flows nearly in the direction of the stress at all depths.

The assumption of an eddy viscosity that is independent of depth is not valid, however, if the water is shallow, because the eddy viscosity must be very small near the bottom, regardless of the character of the current. The effect of a decrease of the eddy viscosity toward the bottom is generally that the angle between wind and current becomes greater at all depths and that the current velocities become greater. This is illustrated by Sverdrup's measurements of wind currents on the North Shelf in latitude 76°35′N, longitude 138°24′E, where the depth to the bottom was 22 m. From these observations, Fjeldstad found that the eddy viscosity could be represented by the formula

$$\mu_e = 385\left(\frac{z + 0.1}{22.1}\right)^{3/4},$$

where the distance z from the bottom is in meters. Fig. 31 shows the observed velocities and the corresponding velocities computed with Fjeldstad's equations and with Ekman's equations, assuming a constant value of μ_e equal to 200.

The question of the time needed for establishing the assumed stationary conditions has also been examined by Ekman, who has made use of a solution given by Fredholm for the case in which a wind suddenly begins to blow with a velocity that later remains constant. It is found that the motion will asymptotically approach a steady state. At each depth the end points of the velocity vectors, when represented as a function of time, will describe a spiral around the end point of the final velocity vector, the period of oscillation being 12 pendulum hours, corresponding to the period of oscillation by inertia movement (p. 94). The *average* velocity over 24 hours will be practically stationary from the very beginning, but the oscillations around the mean motion may continue for several days and may appear as damped motion in the circle of inertia.

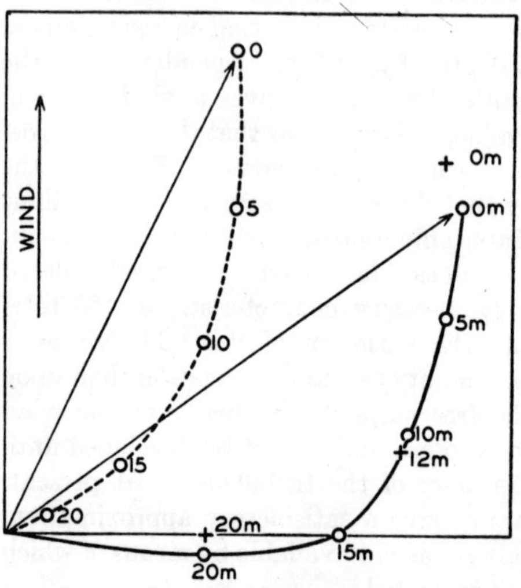

Fig. 31. Wind current in shallow water, assuming a constant eddy viscosity (dashed curve) or an eddy viscosity that decreases toward the bottom (full-drawn curve). Observed currents are indicated by crosses.

Wind Currents in Water in which the Density Increases with Depth

In the equations of motion that govern the wind current, the density enters explicitly, and it might therefore be expected that variations of density in a vertical direction would modify the results, but the variations of the density in the ocean are too small to be of importance in this respect. Indirectly, the variations of density do greatly modify the wind current, however, by influencing the eddy viscosity of the water.

The rate at which a wind current penetrates toward greater depths will depend upon the change of density with depth. Where already an upper, nearly homogeneous layer of considerable thickness has been developed by cooling from above and resulting vertical convection, the wind current will in a short time reach its normal state. Where a light surface layer has been developed because of heating or, in coastal areas, because of addition of fresh water, the wind current will first stir up the top layer and by mixing processes create a homogeneous top layer. When this is formed, the stability will be great at the lower boundary of the layer, where the eddy viscosity will be small, and a further increase of the thickness of the homogeneous top layer will be effectively impeded, although the thickness may be much less than that of the layer within which normally a wind current should be developed. The further increase must be very slow, but no estimate can be given of the time required for the wind current to penetrate to the depths that it would have reached in homogeneous water. The gradual increase of the homogeneous top layer is shown schematically in fig. 32A.

If the wind dies off, heating at the surface may again decrease the density near the surface, but, as soon as the wind again starts to blow,

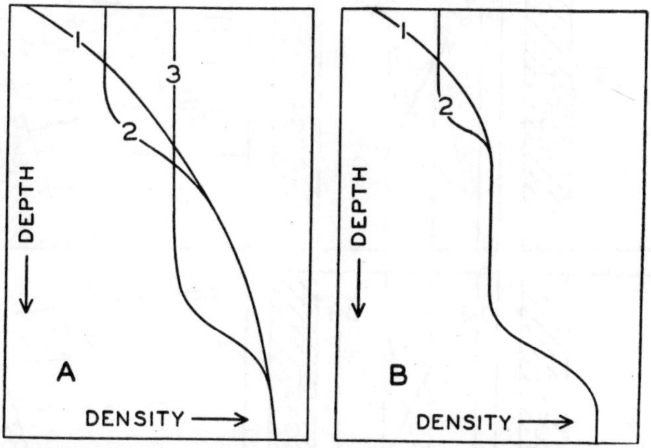

Fig. 32. Effect of wind in producing a homogeneous surface layer demonstrated by showing progressive stages of mixing.

a new homogeneous layer is formed near the surface, and, consequently, two sharp bends in the density curves may be present (fig. 32B).

If conclusions are to be drawn from the thickness of the upper homogeneous layer as to the depth to which wind currents penetrate, cases must be examined in which the wind has blown for a long time from the same direction and with nearly uniform velocity. In middle and high latitudes the wind current will reach its final state more rapidly in winter, when cooling takes place at the surface, than in summer, when heating takes place.

Secondary Effect of Wind in Producing Ocean Currents

In the open ocean the total transport by wind is equal to τ_a/λ and is directed normal to the wind regardless of the depth to which the wind current reaches and regardless of the variation with depth of the eddy viscosity. This fact is of the greatest importance, because the transport of surface layers by wind plays a prominent part in the generation and maintenance of ocean currents. Boundary conditions and converging or diverging wind systems in certain regions must lead to an accumulation of light surface waters, and in other regions to a rise of denser water from subsurface depths. Thus the wind currents lead to an altered distribution of mass, and consequently to an altered distribution of pressure that can exist only in the presence of relative currents. These processes will be illustrated by a few examples.

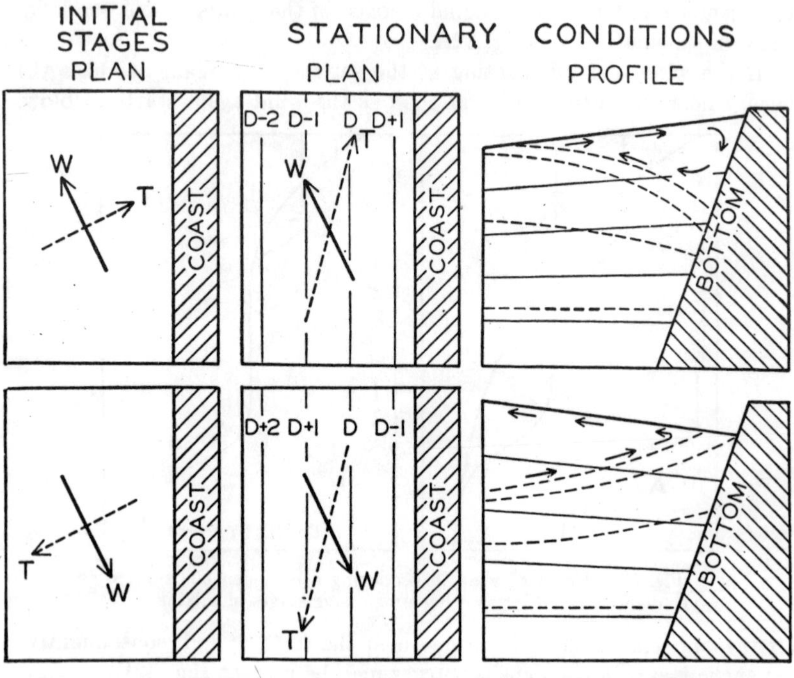

Fig. 33. Schematic representation of effect of wind toward producing currents parallel to a coast in the Northern Hemisphere and vertical circulation. W shows wind direction and T shows direction of transport. Contours of sea surface shown by lines marked D, $D + 1$. . . . Top figures show sinking near the coast, bottom figures show upwelling.

Consider in the Northern Hemisphere a coast line along which a wind blows in such a manner that the coast is on the right-hand side of an observer who looks in the direction of the wind (fig. 33). At some distance from the coast the surface water will be transported to the right of the wind, but at the coast all motion must be parallel to the coast line. Consequently, the convergence that must be present off the coast leads to an accumulation of light water along the coast. This accumulation creates an internal field of pressure with which must be associated a current running parallel to the coast in the direction of the wind. Thus, the wind produces not only a pure wind current but also a *"relative" current* that runs in the direction of the wind. It has not been possible as yet to examine theoretically the velocity distribution within this relative current, but it is *a priori* probable that this current becomes more and more prominent the longer the wind blows. As it increases in speed, eddies will probably develop, and these may limit the velocities that can be attained under any given circumstances. Also, it is probable that a vertical circulation will be present that will consume energy and tend to limit the velocities which can be reached.

Consider next a wind in the Northern Hemisphere which blows parallel to the coast, with the coast on the left-hand side. In this case the light surface water will be transported away from the coast and, owing to the continuity of the system, must be replaced near the coast by heavier subsurface water. This process is known as *upwelling*, and is a conspicuous phenomenon along the coasts of northwest and southwest Africa, California, and Peru. The upwelling also leads to changes in the distribution of mass, but now the denser upwelled water accumulates along the coast and the light surface water is transported away from the coast. This distribution of mass will again give rise to a current that flows in the direction of the wind.

The qualitative explanation of the upwelling was first suggested by Thorade and was developed by McEwen. Recent investigations have added to the knowledge of the phenomenon and have especially shown that water is not drawn to the surface from depths exceeding 200 to 300 m. Deep water does not rise to the surface, but an overturning of the upper layer takes place.

In spite of the added knowledge, it is as yet not possible to discuss quantitatively the process of upwelling or to predict theoretically the velocity and width of the coastal current that develops. It is probable that the forced vertical circulation and the tendency of the current to break up in eddies limit the development of the current. Also, the wind that causes the upwelling as a rule does not blow with a steady velocity, and variations of the wind may greatly further the formation of eddies.

Fig. 34 demonstrates the effect of winds from different directions on the currents off the coast of southern California in 1938. The charts show the geopotential topography of the sea surface relative to the 500-decibar surface, which, in this case, can be considered as nearly coinciding with a level surface. Arrows have been entered on the isolines, indicating that these are approximately stream lines of the surface currents.

In the absence of wind, one should expect a flow to the south or southeast, more or less parallel to the coast. In February, 1938, winds from the south or southwest had prevailed. The light surface water had been carried toward the coast, and, consequently, a coastal current running north was present and was separated from the general flow to the south by a trough line that probably represented a line of divergence. This inshore current to the north may not have been an effect of the wind only, however, but may have represented a countercurrent that developed when variable winds blew (p. 204).

In June, northwesterly winds had prevailed for several months, carrying the light surface water out to a distance of about 150 km from the coast, where the swift current followed the boundary between the light offshore water and the denser upwelled water. Within both types of water several eddies appeared.

Similar considerations apply to conditions in the open ocean. Take the case of a stationary anticyclone in the Northern Hemisphere. Over

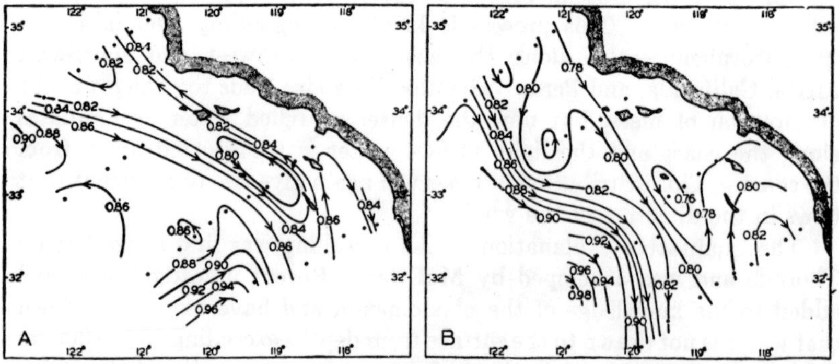

Fig. 34. A. Geopotential topography of the sea surface off southern California relative to the 500-decibar surface in February, 1938, after a period with westerly or southwesterly variable winds. B. Topography and corresponding currents in June after a period with prevailing northwesterly winds.

the ocean the direction of the wind deviates but little from the direction of the isobars, so that the total transport of the wind will be nearly toward the center of the anticyclone. Light surface water will therefore accumulate near the center of the anticyclone and a distribution of mass will be created which will give rise to a current in the direction of the wind. Similarly, the surface water will be transported away from the center of a cyclone, at which heavier water from subsurface depths must rise. Again a field of mass is created, and associated with it will be a current running in the direction of the wind.

Mention has so far been made only of the direction of the wind, but the total transport depends also upon the square of the wind velocity and upon the latitude. In order to find the actual convergences due to wind, it is therefore necessary to take into account both the velocity and the latitude. An attempt in this direction has been made by Montgomery, who finds that in the North Atlantic Ocean the region of maximum convergence lies to the south of the anticyclone.

Every wind system, whether stationary or moving, will create currents associated with the redistribution of mass due to wind transport. It is furthermore possible that within a moving wind system the distribution of mass does not become adjusted to the wind conditions, and that actual piling up or removal of mass may occur such as takes place in partly landlocked seas. If this is true, slope currents (p. 108) reaching from the surface to the bottom develop, but they are of a local character and are soon dissipated. One may thus expect that superimposed on the general currents will be irregular currents due to changing winds and, furthermore, eddies that are characteristic of the currents themselves and independent of wind action. A synoptic picture of the actual currents can therefore be expected to be highly complicated.

Origin of Wind Waves

It is evident to the most casual observer that surface waves are created by wind, but only recently a successful physical explanation of the process has been presented by H. Jeffreys. Jeffreys avails himself of the fact that in a turbulent flow of air, eddies are formed on the lee side of obstacles. Thus, when the wind blows over a sequence of waves, eddies will be formed on the lee side of the waves, for which reason the pressure of the wind will be greater on the windward slopes than on the slopes that are sheltered by the crests. This condition can prevail, however, only if the waves travel at a velocity which is smaller than the speed of the wind. On the basis of these arguments, Jeffreys finds that waves may increase only if

$$c(W - c)^2 \geq \frac{4\nu g(\rho - \rho')}{s\rho'}. \qquad \text{(VII, 23)}$$

Here W is the velocity of the wind, c is the velocity of the waves, ν is the kinematic viscosity of the water, g is the acceleration of gravity, ρ and ρ' are the densities of the water and the air, respectively, and s is a nondimensional numerical coefficient that Jeffreys calls the "sheltering coefficient." It should be observed that in his reasoning Jeffreys takes into account both the turbulent character of the wind and the viscosity of the water. His theory therefore must be expected to give results in better agreement with actual conditions than earlier theories based on the concepts of classical hydrodynamics, which neglect turbulence and viscosity.

The term on the right-hand side of the equation (VII, 23) is always positive. The product on the left-hand side must therefore always be positive and can exceed the right-hand term only if the wave velocity differs sufficiently both from zero and from the wind velocity. For any given wind velocity, there can be only a limited range of possible wave velocities. At a given wind velocity the right-hand side of (VII, 23) will be at a maximum when $c = \frac{1}{3}W$. Therefore, unless

$$W^3 \geq \frac{27\nu g(\rho - \rho')}{s\rho'}, \tag{VII, 24}$$

there will be no values of c that satisfy condition (VII, 23).

Equation (VII, 24) determines the velocity of the weakest wind capable of raising waves, and this wind could be determined if the sheltering coefficient were known. Jeffreys has not been able to make independent determinations of this coefficient, but has, instead, conducted wind measurements over ponds in order to determine the lowest velocity at which small waves appear. He found that at wind velocities of less than 1 m/sec no disturbance of the surface occurred, but that at a velocity of about 1.1 m/sec distinct waves appeared. The corresponding value of the sheltering coefficient, s, would be about 0.27. It should be observed, however, that, because of the rapid change with height of wind velocity near a boundary surface, this numerical value and the limiting wind velocity both depend upon the height above the water at which the wind velocity was measured.

The velocity of the smallest possible waves should be one third of the limiting wind velocity, or about 37 cm/sec, and according to the theory (p. 135) the corresponding wave length must be 8.8 cm. Thus, measurement of the smallest waves can be used for testing the correctness of the theory, but measurement of such small wave lengths is very difficult, and no exact observations have been made. The length of the shortest waves observed by Jeffreys lies in the neighborhood of the theoretical value, but the shortest waves measured by Scott Russell had a length of 5 cm, and those measured by Cornish were only about 2.5 cm long. The problem of the generation of surface waves is therefore not satisfactorily solved, but Jeffreys' approach is in better agreement with observations than any made previously.

It should be added that, if only the forces due to surface tension and gravity are considered, waves should not be formed until the wind velocity passes the limit of 6.7 m/sec and, if only the stress of the wind on the surface and gravity are taken into account, the limiting wind velocity would be about 4.3 m/sec. Experience shows that these values are far too high, and the turbulent character of the wind must therefore be of the greatest importance.

Form and Characteristics of Wind Waves

In physics the general picture of surface waves is that of sequences of rhythmic rise and fall. Progressive waves appear to move along the surface, and standing waves appear stationary. The actual appearance of the surface of the open sea, however, is mostly in the sharpest contrast to that of rhythmic regularity. If a wind blows, waves of many different sizes are present, varying in form from long, gently sloping ridges to waves of short and sharp crests. Superimposed on the gentler waves, which may or may not run in the direction of the wind, appear series of deformations of the surface which, from the point of view of physics, can be termed "waves" only by stretching the definition.

In spite of the irregular appearance of the sea, it is possible to apply the terms *wave period*, T, *wave height*, H, and *wave length*, L, because some of the waves will be more conspicuous than others and their characteristics can be observed. The general theory of waves on the surface of the sea leads to a simple formula for wave velocity:

$$c = \frac{L}{T} = \sqrt{g\frac{L}{2\pi}} = g\frac{T}{2\pi}, \tag{VII, 25}$$

from which the relations

$$L = \frac{2\pi}{g}c^2 = \frac{g}{2\pi}T^2 \tag{VII, 26}$$

and

$$T = \sqrt{\frac{2\pi L}{g}} = \frac{2\pi}{g}c \tag{VII, 27}$$

are derived. These formulas apply only to a wave whose amplitude is small relative to the length, and therefore they cannot be expected to be valid in all cases.

Of the three interrelated quantities, c, T, and L, the wave period T can probably be most easily determined at sea by using the method proposed by Cornish, which consists in recording the time intervals between appearances of a well-defined patch of foam at a sufficient distance from the ship. The same method can be used on the coast, where, in addition, the interval between breakers can be accurately timed. At sea the wave length is mostly estimated on the basis of the ship's length, but this procedure leads to uncertain results because it is often difficult to locate both crests of the wave relative to the ship, and also because of disturbance due to the waves created by the moving ship. The most satisfactory measurements are made from a ship that is hove to. Another method consists in letting out a floater as a wave crest passes the stern of a ship and recording the length of line paid out when the floater reappears on the crest aft of the ship, as well as the angle that the line forms with the direction in which the ship travels. The velocity of the wave can be found by recording the time in which the wave runs a measured distance along the ship. If the period is also determined, the wave length is found from the simple formula $L = cT$.

A large number of measurements have been made at sea in order to establish the relationship between the period of wave, the wave length, and the wave velocity. Critical examination of the methods employed has been made, especially by Cornish, and a number of average results have been compiled by different authors. Table 15 contains one of the compilations made by Krümmel, from which it is evident that the observed values are in fair agreement with the theoretical expectation.

TABLE 15

OBSERVED AND COMPUTED VALUES OF VELOCITIES, LENGTHS, AND PERIODS OF SURFACE WAVES

Region	Wave velocity, m/sec			Wave length, m			Wave period, sec		
	Observed	Computed from $\sqrt{g\frac{L}{2\pi}}$	Computed from $g\frac{T}{2\pi}$	Observed	Computed from $\frac{2\pi c^2}{g}$	Computed from $g\frac{T^2}{2\pi}$	Observed	Computed from $\sqrt{\frac{2\pi L}{g}}$	Computed from $\frac{2\pi c}{g}$
Atlantic Ocean Trade wind region	11.2	10.8	10.5	65	70	61	5.8	6.0	6.2
Indian Ocean Trade wind region	12.6	13.1	13.7	96	88	104	7.6	7.3	6.9
South Atlantic Ocean West wind region	14.0	15.5	17.1	133	109	163	9.5	8.6	7.8
Indian Ocean West wind region	15.0	15.2	13.7	114	125	104	7.6	8.0	8.3
China Sea	11.4	11.9	12.4	79	72	86	6.9	6.6	6.3
Western Pacific Ocean	12.4	13.6	14.7	102	85	121	8.2	7.5	6.9

The longest wave periods observed at sea rarely exceed the value of 13.5 sec that Cornish reports from the Bay of Biscay. It has been established, however, that the swell reaching the shore may have much longer periods and correspondingly longer wave lengths and greater velocities of progress. The longest period that Cornish has observed is about 22.5 sec, corresponding to a wave length in deep water of about 850 m and a velocity of progress of 35 m/sec. This difference between the waves of the open ocean and the swells that reach the coast will be dealt with later.

The theory of waves leads not only to a relation between the wave length and the velocity of progress, but also to results concerning the profile of the wave. These results are based on the concepts of classical hydrodynamics and have been derived from the hydrodynamic equations, omitting friction but taking the boundary conditions into consideration. In an ideal fluid the free surface of the waves, according to Stokes, will very nearly take the shape of a trochoid—that is, a curve that is formed by the motion of a point on a disk when this disk rolls along a level surface. If the amplitude is small, the trochoid approaches in shape a sine curve, but at great amplitude the crests become narrower and the troughs longer.

Stokes' results lead to the conclusion that at increasing amplitude the wave form deviates more and more from the trochoid. Studies of the stability of waves by Michell show that the wave must become unstable if the angle formed by the crest is less than 120 degrees. In this case, the ratio of height to length is 1:7 (see fig. 35). The velocity of progress of these waves is no longer independent of the height, but can be written

$$c = \left[g\frac{L}{2\pi}\left(1 + \pi^2\left(\frac{H}{L}\right)^2\right)\right]^{1/2}$$

In the case of the extreme Michell wave the velocity of progress is about 1.14 times greater than that of waves of small amplitude.

Accurate measurements of actual wave profiles would be very useful for examining the correctness of the above-mentioned theoretical conclusions. Such measurements have been based on photogrammetric

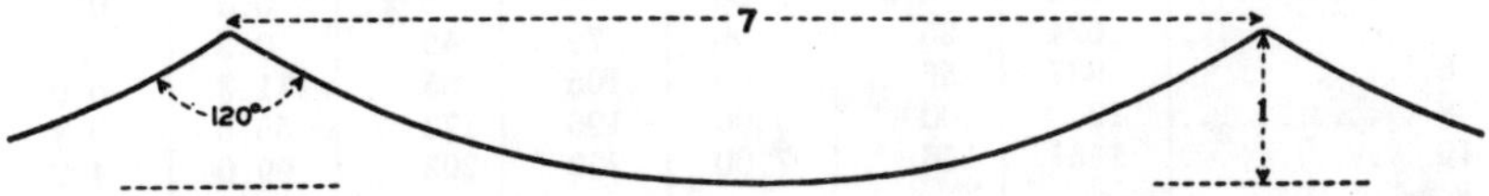

Fig. 35. True dimensions of steepest possible wave, according to Stokes and Michell.

pictures, but the observed profiles show little similarity to the theoretical curves.

The wave theory also leads to certain conclusions concerning the character of motion of individual water particles. In a wave that has the form of a trochoid, the single water particles will describe circles whose radii decrease with increasing depth:

$$r = ae^{-2\pi\frac{z}{L}}, \qquad \text{(VII, 28)}$$

where a is the amplitude of the wave ($a = \frac{1}{2}H$), L is the wave length, and z is the depth below the undisturbed water surface. Each water particle describes a circle with radius r in the time T that represents the period of the wave. The velocities of the individual water particles are, then,

$$v = \frac{2\pi}{T}ae^{-2\pi\frac{z}{L}}. \qquad \text{(VII, 29)}$$

These formulas are valid only when the amplitude of the wave is small relative to the length, but they can nevertheless be used for an approximate computation of the greatest velocities that may be encountered in waves. The first columns of table 16 show the periods and the corresponding wave length and velocities of progress and the assumed values of the heights of waves. It should be observed that the height equals twice the amplitude. The heights entered in the table reach approximately the greatest observed heights of waves of stated periods up to 14 sec, and the last three lines of the table correspond to big swells. The last four columns of the table give the velocities of the water at the surface and at the depths of 2, 20, and 100 m. It is seen that the surface

TABLE 16

VELOCITIES OF WATER PARTICLES AT DIFFERENT DEPTHS IN SURFACE WAVES OF DIFFERENT PERIODS, LENGTHS, AND HEIGHTS

Wave characteristics				Velocity of particles in cm/sec at stated depths			
Period length (sec)	Velocity of progress (cm/sec)	Length (m)	Height (m)	0 m	2 m	20 m	100 m
2	312	6.2	0.25	39	5.2	0.0	0.0
4	624	25	1.00	79	49	0.5	0.0
6	937	56	2.00	105	85	11.3	0.0
8	1249	100	5.00	196	173	55.6	0.4
10	1561	156	7.00	220	203	99.0	4.2
12	1873	225	10.00	211	199	114.0	12.9
14	2185	306	12.00	273	262	180.0	35.0
16	2498	396	10.00	197	190	143.0	40.6
18	2810	506	8.00	140	136	109.0	40.5
20	3122	624	5.00	78	76	63.0	28.4

velocities can reach very appreciable values, up to 250 cm/sec or more, but in the case of the shorter waves the velocity decreases very rapidly with depth and is negligible shortly below 20 m. In waves of periods below 10 sec the wave motion is negligible below 100 m. The tabulated velocities correspond to the heights that are entered in the table, and at different wave heights the velocity will be altered proportionately.

No measurements are available of the actual motion of water particles in waves. Experience in submarines has shown, however, that the wave motion decreases rapidly with increasing depth. Vening Meinesz has availed himself of this fact and has been able to conduct observations of gravity at sea on board a submarine, making use of pendulums, which can be employed only when the motion of the vessel is small.

A consequence of the decrease of the particle velocity with depth is that a small transport of water will take place in the direction of progress. A water particle will move in the direction of progress when it is above its mean depth, and in the opposite direction when it is below its mean depth; owing to the decrease of velocity with depth, it will move somewhat faster in the direction of progress than in the opposite direction. Consequently, after completing one revolution in its orbit, the particle will not return to the point from which it started, but it has advanced somewhat in the direction of progress, meaning that an actual transport of water takes place in this direction even in the absence of wind.

The irregular appearance of the surface is not accounted for by the theories that have been mentioned so far, but a somewhat better understanding of the pattern of waves is obtained when one takes into account the phenomena of interference. Suppose that two waves travel in the same direction but with a slightly different velocity. The waves at the surface can then be represented by the equations

$$\xi_1 = a_1 \sin(\kappa_1 x - \sigma_1 t) \quad \text{and} \quad \xi_2 = a_2 \sin(\kappa_2 x - \sigma_2 t), \qquad \text{(VII, 30)}$$

where ξ is the vertical displacement of the sea surface, where $\kappa = 2\pi/L$, and where $\sigma = 2\pi/T$. The actual appearance of the surface is obtained by adding the displacements due to the two individual waves. If the amplitudes are equal, one obtains

$$\xi = 2a \cos[\tfrac{1}{2}(\kappa_1 - \kappa_2)x - \tfrac{1}{2}(\sigma_1 - \sigma_2)t] \sin[\tfrac{1}{2}(\kappa_1 + \kappa_2)x - \tfrac{1}{2}(\sigma_1 + \sigma_2)t]. \qquad \text{(VII, 31)}$$

If the wave lengths differ by a small amount only, this new equation represents a wave whose length is the average of the two waves of which it is composed but whose amplitude varies between zero and $2a$. At the locality where the two waves are in phase, the amplitudes are added, and a wave appears of twice the amplitude of the two original waves, but, where the waves are in opposite phases, the amplitudes cancel. The free surface takes the appearance of a sequence of wave groups separated by regions with practically no waves. A simple pattern of interference is involved, and, owing to this interference, the two individual waves are no longer conspicuous, but are replaced by a series of wave trains that appear to progress with a definite velocity:

$$c = \frac{\sigma_1 - \sigma_2}{\kappa_1 - \kappa_2}. \qquad \text{(VII, 32)}$$

If the wave lengths are only slightly different, this velocity is very nearly equal to $\frac{1}{2}c$, where c now represents the average velocity of the two interfering waves. Thus, the wave train progresses with a velocity that is only one half that of the single waves, which therefore advance through the wave trains.

The upper curve in fig. 36 is a reproduction of a record of waves obtained at the end of the Scripps Institution of Oceanography pier and represents a good example of interference of waves of nearly the same period length but of different amplitudes. In the middle portion of the figure are shown two sine curves, one of period 9.6 sec and amplitude 0.75 m, and one of period 8.7 sec and amplitude 0.32 m. The heavy curve represents the wave pattern that would result from interference between the two, and it is seen that this roughly corresponds to the observed pattern. The discrepancies are accounted for partly by the fact that the record was obtained only about 300 m from the beach, where the depth to the bottom was about 6 m, for which reason the waves were somewhat deformed, and partly by the fact that waves of shorter periods apparently were present. Such relatively clear-cut cases are rare because, for the most part, waves of so many different period lengths are present that the resulting pattern of interfering waves is extremely complicated. The lower curve in the figure reproduces a record which

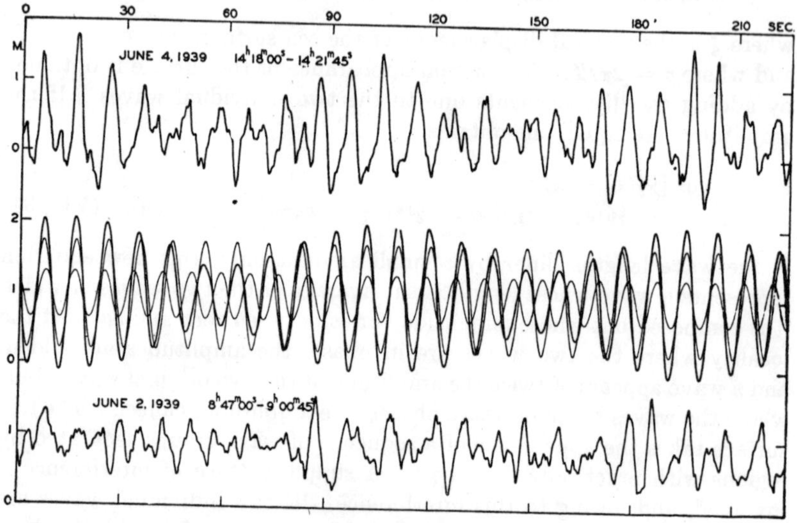

Fig. 36. *Upper curve:* Record of waves at the end of the Scripps Institution pier, showing interference. *Middle curve:* Computed pattern of wave interference. *Lower curve:* Example of the ordinary type of records of waves at the Scripps Institution pier, showing very complicated conditions.

has been selected at random and which shows isolated high waves occurring at apparently irregular intervals.

These considerations help to explain the occurrence of a sequence of high waves followed by a sequence of low ones, but they do not explain the irregular pattern of "waves" that is called "cross sea." Progress toward explaining the typical cross sea, however, has recently been made by H. Jeffreys. Jeffreys calls the waves that have been discussed so far *long-crested* waves, because it has been assumed that the crest of the wave is very long compared to the wave length. Mathematically speaking, it is assumed that the wave crests are of infinite length. Jeffreys had introduced the so-called *short-crested* waves, which can be represented by the formula

$$\xi = a \cos (\kappa x - \sigma t) \cos \kappa' y,$$

where $L' = 2\pi/\kappa'$ represents the length of the crest. This formula defines a series of waves that travel at a speed somewhat greater than that of the long-crested waves, the crest itself forming a wave line at right angles to the direction of progress: $c' = c(1 + L^2/L'^2)^{1/2}$ (fig. 37). Since such a wave can also travel without altering form, it therefore belongs to the group of waves which are theoretically possible. It is interesting to observe that, according to Jeffreys, the first waves which are generated by the weakest winds must be long-crested, but, when the wind increases in velocity, short-crested waves can be formed. The explanation is that the turbulence of the wind is characterized not only by random motion in the direction of the wind, but also by random motion at right angles to the wind. At higher wind speeds, the turbulent velocities at right angles to the wind may be great enough to break the originally long-crested waves and to create new short-crested waves. These results help greatly toward understanding the irregular appearance of the waves in strong winds.

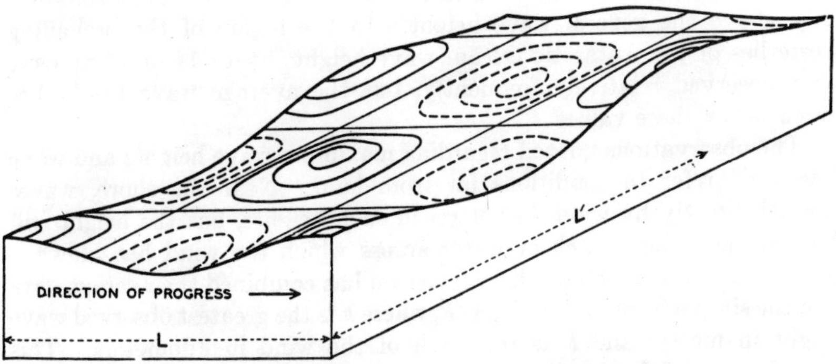

Fig. 37. Schematic picture of short-crested waves.

In table 15, p. 136, only wave velocity, wave length, and wave period are listed, and no information is given as to the wave height. A large number of measurements of wave heights have been made by different methods which, although uncertain, are considered more accurate than the measurements of wave lengths. The wave height can be found if a location on board ship can be selected at which the tops of the waves appear level with the horizon. The wave height is then equal to the eye height of the observer above the water line. Another method is based on the records of a delicate barometer, and still another that gives very accurate results makes use of photogrammetric measurements.

The greatest wave heights observed in most oceans are about 12 m. Cornish gives a very vivid description of waves of that height during a storm which he experienced in the Bay of Biscay in December, 1911.

On the day preceding the gale a heavy swell with a period of 11.4 sec and an average height of about 6 m came from the northwest. The wind, which had blown as a breeze from the southwest, changed during the night to west-northwest and increased in the morning to a strong gale with velocities up to 23 m/sec. The period of the waves increased to 13.5 sec, corresponding to a length of 310 m and a velocity of progress of 21 m/sec, while the wave height increased up to 12 m. Accounts of similarly large waves are given in many cases. In a hurricane in the North Atlantic in December, 1922, when the wind velocity probably passed 45 m/sec, one of the officers of the *Majestic* reported waves that averaged more than 20 m in height and reached a maximum height of up to 30 m. It is probable, however, that the greatest wave heights refer to occasional peaks of water which may shoot up to elevations considerably above the general wave height. In the region of the prevailing westerlies of the Antarctic Ocean, wave heights up to 14 or 15 m have been observed relatively frequently, but the average wave height lies much below these values.

The observations quoted regarding maximum wave heights and wave length all refer to conditions far from land. Near the shore, waves created directly by wind do not reach such heights, but the height will depend upon the stretch of water across which the wind has blown—that is, the fetch of the wind. Stevenson has combined the average data into the simple formula $h = \frac{1}{3}\sqrt{F}$, where h is the greatest observed wave height in meters and F is the fetch of the wind in kilometers. This formula is valid for small bodies of water and is also applicable up to a certain distance from the coast when the wind blows away from the coast. If the fetch of the wind is less than about 10 km, a small correction term has to be added. Not only the wave height but also the wave length increases with increasing distance from shore.

The ratio between the length and the height of the waves appears to vary between 10 and 20 when a fresh wind blows, but, in the case of a swell, may lie between 30 and 100. Results that have been obtained by different observers will be mentioned later (p. 145).

Relations Between Wind Velocity and Waves

In order to explain some of the phenomena thus far discussed, it is necessary to consider the energy of the waves. This energy can be computed by considering that it is present partly as potential energy and partly as kinetic energy. The computation leads to the result that in the case of long-crested waves the energy per unit area of the sea surface is approximately equal to $\frac{1}{2}\,g\rho a^2$, where g is the acceleration of gravity, ρ is the density of the water, and a is the amplitude of the wave. In the case of short-crested waves the energy per unit area of sea surface is approximately one half of this amount.

The energy of the waves is transmitted to them by the wind, and, according to Jeffreys, the processes that lead to the generation of waves are also of fundamental importance to their further development. When the wind velocity increases, the pressure exerted on the windward side of the wave will be greater than that on the lee side. Consequently, the waves will increase in height and their energy will increase also. The energy considerations do not lead to any limit of height to which the waves can grow, but the wave theory itself, as developed by Michell (see p. 137), shows that the height cannot exceed about one seventh of the wave length. Thus, if the wind constantly imparts energy to the waves, and if there is no increase in the wave length, the crests will merely break and the sea will become covered by whitecaps. However, the longer the wave the greater the height it can reach, and thus the greater the amount of energy it can absorb from the wind. The wind is able to produce waves of different length, but the longer will continue to grow for a longer time. Jeffreys therefore concludes: "When the waves have traveled a long distance, with the wind blowing them all the time, the longest waves will tend to predominate, simply because they can store more energy."

However, there is also a limit to the length that waves can attain, because the velocity of progress of the waves increases with increasing wave length and because the wind cannot impart further energy to the waves if they travel at a speed that is greater than the wind velocity. At a given wind velocity the longest possible waves will therefore be those that travel at a velocity somewhat below the wind velocity. According to Cornish the speed of the fully developed waves, on an average, is eight tenths of the speed of the wind, but it should be borne in mind that the observed wind velocity depends upon the height at which the velocity is measured. Information is still lacking as to the relation between the velocity of the wind directly over the sea surface and the velocity of the waves.

It also follows that the greatest wave velocity cannot exceed the velocity of the wind that has created the waves, assuming that a surface wave continues at a constant speed after leaving a region of strong winds. Theories developed by Poisson and Cauchy, however, lead to the conclusion that the wave length—and therefore the velocity of progress and the period—of a surface wave increases in course of time. The velocity of the swell reaching the coast, according to these theories, should be greater than the velocity of the waves that were directly created by the wind. Krümmel quotes a single observation that may support these conclusions, whereas Cornish emphasizes the fact that no breakers have been observed whose velocity of progress in deep water exceeded observed wind velocities. In the case of breakers with a period of 20 sec, corresponding to a velocity of progress of about 30 m/sec, which were observed on the coast of the English Channel on December 29, 1898, Cornish points out that prior to the arrival of these breakers a gale had been reported in mid-Atlantic in which the force of the wind had probably exceeded 35 m/sec, which was about 5 m/sec greater than the speed of the waves. He considers it probable that these waves were formed within the area of the gale and traveled for a long distance at their original speed. Waves may nevertheless reach the coast before the arrival of the gale, because the speed at which an atmospheric disturbance travels is, in general, less than the maximum speed of the wind within the disturbance.

Cornish also points out that in the open ocean the existence of long swells can be completely obscured if shorter waves of greater height are present at the same time. He illustrates his point by means of a graph similar to the one shown in fig. 38. It is here assumed that two waves are present, one long swell, curve A, and one wave, curve B, which has only one third the period of the swell but twice as great an amplitude.

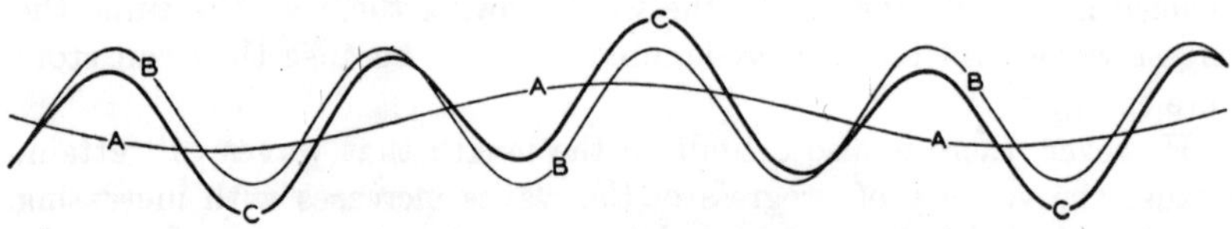

Fig. 38. Interference between a long swell (A) and a much shorter wave (B).

By the interference of these two waves the surface takes the appearance shown by the heavy curve, C. In this case an observer would get the impression that the waves present were of the same period as the shortest waves, but of variable height, a phenomenon that is often recorded. The long swell therefore may very well be obscured by shorter waves.

A number of studies have dealt with the relation between the wind velocity and the maximum wave height, the steepness, and the velocity of progress of the waves. For the correlation of quantities that have not been directly measured, the equations on p. 135 give the theoretical relations between wave velocity and wave length or wave velocity and wave period.

Cornish's empirical results regarding the highest waves can be summarized as follows:

$$c = 0.8W \qquad \text{and} \qquad \frac{L}{H} = \frac{5}{3}T.$$

By means of the equations on p. 135, one obtains the relation $H = 0.48W$, indicating that the highest wave heights are proportional to the wind velocity. Zimmerman's empirical results are

$$L = 10.62H^{4/3} = 3.55W^{4/3},$$

giving $H = 0.44W$, in good agreement with the results of Cornish as to relation between wave height and wind. Krümmel, on the other hand, arrives at the conclusion that the maximum wave heights are greater than those corresponding to a linear relationship, and Rossby has for theoretical reasons suggested a formula of the type

$$H = \frac{G}{g} W^2,$$

where G is a nondimensional constant. When $G = 0.3$, Rossby finds that his formula fits the available data fairly well at high wind velocities, but it may be stated nevertheless that the relation between wind velocity and wave height is so complicated that nc simple empirical formula has so far been established.

The same complication exists when the steepness of the waves is expressed by means of the ratio between wave length and wave height, L/H, which is inversely proportional to the steepness. Cornish's relations lead to the formula $L/H = 0.85W$, but this is evidently not valid at wind velocities much below 10 m/sec, because the steepest possible waves are characterized by a ratio $L/H = 7$. Zimmerman's values give $L/H = 8.1W^{1/3}$. These two results are in qualitative agreement, because both indicate an increase of the ratio L/H with increasing wind velocity. They are also in agreement with Jeffreys' explanation of the growth of waves, according to which one must expect L/H to be greatest for the longest waves. Schott, on the other hand, found that the ratio L/H decreased with increasing wind velocity, and, in discussing a large number of observations by Paris, he found the ratio to be constant. Observations from lakes indicate that there the ratio varies between 10 and 12.

A similar confusion exists regarding the relation between wave velocity and wind velocity. Cornish found, as already stated, that $c = 0.8W$, but from Zimmerman's relation it follows that $c = 2.36W^{2/3}$. According to the latter equation the wave velocity is greater than that of the wind up to a wind velocity of 13.2 m/sec, and smaller when the wind is above that value, in disagreement with energy considerations, which lead to the conclusion that the wave velocity must always be less than the wind velocity.

All these discrepancies indicate that wave height, wave profile, and velocity of progress are not dependent upon the wind velocity at the time of observation alone, but may also depend upon the length of time the wind has blown, the state of the sea when the wind started blowing, and the dimensions of the area over which the wind has blown. It can in all events be stated that comprehensive observations are needed for clearing up these questions.

It was mentioned (p. 142) that the wave height depends upon the fetch of the wind, and that for small bodies of water a simple empirical relationship had been established between maximum wave heights and the dimensions of the body. This formula is valid to a distance of 600 to 900 nautical miles, at which the maximum wave height characteristic of the open ocean, about 12 m, may be reached. The fact that greater wave heights are rarely observed may be due to the circumstance that the wind systems usually have dimensions smaller than 1000 nautical miles, so that in the open ocean the actual fetch of the wind cannot exceed that distance, or it may be that a longer fetch of the wind tends more toward increasing the length of the waves than toward increasing their height.

The difference in energy of long-crested and short-crested waves and of high and low waves must be considered in the discussion of what happens to the waves when the wind stops blowing. Owing to their smaller energy per unit area the short-crested waves will be destroyed more rapidly by friction, and the largest of the long-crested waves stand the best chance of surviving for longer periods. It is also probable that the dissipation of energy is more rapid within the steeper waves, for which reasons the steeper and shorter waves will be destroyed more rapidly than the longer and less steep. Thus one should expect that, outside of the region in which the wind blows, long-crested swells will become more and more dominating, and, at considerable distances from the wind areas, only long-crested swells will be present. These conclusions are in good agreement with observed conditions.

Waves Near the Coast. Breakers

When the waves approach the coast, a number of things occur. It is conspicuous, first, that the short-crested cross sea mostly disappears at some distance from the coast and that it is mainly the long-crested rollers that reach the beaches. Jeffreys has been able to show that this transformation is associated with the change in form and the dissipation of energy that take place when surface waves enter shallow water. One of the characteristic deformations is that originally symmetrical waves become unsymmetrical when the depth decreases, the front of the waves becomes steeper, and, finally, the waves break. This effect is more conspicuous in the case of the short-crested waves, which therefore break at a greater distance from the coast, whereas the long-crested waves can proceed farther without being destroyed.

Another effect of the decrease of depth is that the wave velocity is reduced. Consequently, a wave that approaches the coast at an angle will be deflected so that it will reach the coast with the wave front nearly parallel to it. The part of the wave that first approaches the coast will be slowed down, but the outer portion of the wave will still advance with a great velocity and the direction of the wave front will be turned.

Still another effect is related to the fact that when approaching a coast the waves take on a character which is intermediate between surface waves and long waves—that is, their lengths become great compared to the depth of the water. The waves of long period and correspondingly great wave length are transformed into long waves at a greater distance from the coast than are those of short period. As a consequence, the movement of the water particles of the longer waves will reach to the bottom at a greater distance from the coast, although the height of these waves may be smaller than that of the short-period waves. This circumstance may have considerable bearing on sand movement caused by waves in shallow water.

The breaking of the wave is partly due to friction, which is effective when the water becomes so shallow that the motion reaches from the surface to the bottom. In this case the lower portion of the wave will be slowed down more than the upper portion and the deformation of the wave will be accelerated.

When a wave breaks near the shore, another wave type, known as a wave of translation, may develop. This wave, which was discovered and studied by Russell, is characterized by having only a crest, and no trough. In such a wave the motion of water particles is only in the direction of progress, and the water particles are therefore displaced forward as the wave passes. It is formed when a mass of water is suddenly added to still water, and may therefore be produced as the crest of a breaking wave topples over and crashes down on the water surface in front. This wave type is unimportant in the open sea, but may be prominent on a shallow coast.

Certain phenomena that appear to be associated with breakers are not yet understood. The existence of undertow has not been satisfactorily explained and is doubted by some observers. The rip currents which flow away from the coast through the breakers and which may carry swimmers far out from the beach are probably associated with the surface transport of water against the beach by the waves.

Destructive Waves

The destructive waves that occasionally inundate low-lying coasts and cause enormous damage are commonly known as "tidal waves," although they have nothing in common with the tides. The name "tidal wave" has, however, become so firmly established in the English language that the popular use will probably be continued in spite of the unfortunate confusion to which it gives rise. Destructive waves are not related to the tide-producing forces, but are caused by earthquakes or by severe storms blowing against the coast. It is therefore necessary to distinguish between earthquake waves and storm waves, since the former are real waves and the latter are not even related to waves (p. 122).

Waves in the sea caused by earthquakes are of two different types. In the first place, a submarine earthquake may produce longitudinal oscillations in the sea which proceed at the velocity of sound waves. When reaching the surface, such a longitudinal oscillation will be felt on board a ship as a shock which violently rocks the vessel. The shock may be so severe that the sailors believe their vessel has struck a rock; on early charts, several such reported "rocks" were indicated in waters where recent soundings have shown that the depth to the bottom is several thousand meters. There are many ship reports dealing with such shock waves, particularly from regions in which seismological records show that submarine earthquakes are frequent. Explosion waves of this character occur mostly as independent phenomena, but occasionally they are accompanied by the release of large amounts of gases that rise toward the surface and, owing to their pressure, may lift the surface up like a dome and produce a transverse wave that will spread out like another gravitational wave. Observations of this kind of waves are rare, but it is possible that ships which have been lost at sea have been completely destroyed by such enormous disturbances. When a wave of this nature spreads out from the place where it reaches the surface, it decreases in amplitude, and by the time that it reaches the coasts it has usually been so much reduced that it does not cause much damage.

Destructive waves caused by earthquakes are in general associated with submarine landslides, which directly create transverse waves. These are called *dislocation waves*, and may reach enormous dimensions both in the open sea and near the coasts. They proceed as ordinary long gravitational waves, and many records exist of such waves that have traversed the entire Pacific or Atlantic Ocean and caused enormous damage by completely inundating low-lying areas. Thus, the great damage caused by the earthquake at Lisbon on November 1, 1755, was due mainly to the gigantic wave that was set up; this wave crossed the Atlantic Ocean and reached the West Indies as a "tidal wave" 4 to 6 m high. In Japan, similar earthquake waves have on many occasions brought great destruction and led to the loss of many lives. As an example, it may be mentioned that in 1703 more than 100,000 persons lost their lives when the coast of Awa was flooded. Some of the most discussed waves are those that accompanied the eruption of the volcano Krakatoa, in the Sunda Strait, on August 26 and 27, 1883. Several waves occurred after the different eruptions, and the highest ones caused great devastation on various East Indian islands, where more than 36,000 persons lost their lives and where the waves in certain localities must have reached a height of up to 35 m. These waves did not enter the Pacific Ocean, but crossed the Indian Ocean and entered the Atlantic Ocean, where they were recorded as far north as to the English Channel, having traveled a distance corresponding to half the circumference of the earth in thirty-two and a half hours. In the English Channel the height had decreased, however, to a few centimeters.

As already stated, these waves proceed as long gravitational waves, and their velocity of progress over a uniform bottom, therefore, should be equal to $\sqrt{gh}$. Where the depth to the bottom, h, is variable, the velocity of progress will be less than $\sqrt{g\bar{h}}$, where $\bar{h}$ is the average depth, but it has been found that the velocity of progress is smaller than should be expected even if variations in depth are considered. The study of the rate of propagation of these waves served, in spite of this circumstance, to give an idea of the average depth of the ocean prior to the time of deep-sea soundings. Thus, in 1856, Bache computed the average depth of the oceans to be about 4000 m, whereas Laplace had assumed an average depth of about 18,000 m.

Destructive "waves" caused by wind and barometric pressure are of an entirely different nature. In this case one has to deal not at all with the effect of a wave, but, instead, with inundations that are almost directly caused by the ocean waters being swept up against the coast by violent storms. Abnormally high-water levels caused by strong winds are frequent on many coasts, but, fortunately, the sea level rarely rises so much that great damage occurs. The most destructive storm "wave" known in the history of the United States is that which practically destroyed Galveston on September 8, 1900. A West Indian hurricane approached the coast of the Gulf of Mexico, where, at Galveston, the barometric pressure fell from 29.42 inches at noon to 28.48 at 8:30 P.M. At the same time the wind velocity increased to 100 miles per hour at about 6:00 P.M., when the anemometer was broken to pieces. It has been estimated that the average wind velocity between 6:00 and 8:00 P.M. must have been about 120 miles per hour. During the day the water rose slowly but steadily until the wind had reached hurricane force, when a much more rapid rise took place. In the evening the water level was nearly 15 feet above mean high water, and large districts of the city were flooded. Nearly 6000 persons were drowned, and the property damage ran into tens of millions of dollars.

The hurricane that on September 21, 1938, struck the coast of New England brought an even higher water level in many localities, but did not cause so much loss of life. At Buzzard's Bay the highest water level ranged from 12 to 15 feet above mean low water, and at Fall River it was reported that "the water came up rapidly in a great surge," rising to about 18 feet above normal. Nearly 600 persons lost their lives, and the property damage has been estimated at $400,000,000, but only part of this damage was due to destructive waves.

It should again be emphasized that destructive "waves" caused by wind and also commonly included among "tidal waves" are not waves, but represent a rise of sea level comparable to the minor changes of sea level that are associated with wind and are known on all coastal areas.

CHAPTER VIII

Thermodynamics of Ocean Currents

The preceding description of the effect of the wind, especially the discussion of the secondary effect of the wind in producing currents in stratified water, may leave the impression that the wind is all-important in the development of the ocean currents and that thermal processes can be entirely neglected. Such an impression, however, would be very misleading. In discussing the secondary effect of the wind, it was repeatedly mentioned that the development of the currents caused by a redistribution of mass by wind transport would be checked partly by mechanical processes and partly by thermal processes. Surface waters that were transported to higher latitudes would be cooled, and thus a limit would be set to the differences in density which could be attained; upwelling water would be heated when approaching the surface, and at a certain vertical velocity a stationary temperature distribution would be established at which the amount of heat absorbed in a unit volume would exactly balance the amounts lost by eddy conduction and by transport of heat through the volume by vertical motion. The establishment of a stationary temperature distribution within upwelling water would check the effect of upwelling on the horizontal distribution of density.

The above examples serve to emphasize the importance of the thermal processes in the development of the currents, but an exact discussion of the thermodynamics of the ocean is by no means possible. So far, the principles of thermodynamics have found very limited application to oceanographic problems, but this does not mean that the thermal processes are unimportant compared to the mechanical.

THERMAL CIRCULATION. The term "thermal circulation" will be understood to mean a circulation that is maintained by adding heat to a system in certain regions and by cooling it in other regions. The character of the thermal circulation in the ocean and the atmosphere has been discussed by V. Bjerknes and collaborators. Their conclusions can be stated as follows: If within a thermal circulation heat shall be transformed into mechanical energy, heating must take place under higher pressure and cooling under lower pressure. Such a thermodynamic machine will run at a constant speed if the mechanical energy that is produced by the thermal circulation equals the energy that is expended in overcoming the friction.

In the ocean, "higher pressure" can generally be replaced by "greater depth," and "lower pressure" by "smaller depth." Applied to the ocean the theorem can be formulated as follows: If within a thermal circulation heat shall be transformed into mechanical energy, the heating must take place at a greater depth than the cooling.

This theorem was demonstrated experimentally by Sandström before it was formulated by Bjerknes. In one experiment, Sandström placed a "heater" at a higher level and a "cooler" at a lower level in a vessel filled with water of uniform temperature (fig. 39a). The heater consisted of a system of tubes through which warm water could be circulated, and the cooler consisted of a similar system through which cold water could be circulated. When warm and cold water was circulated through the pipes, a system of vertical convection currents developed and continued

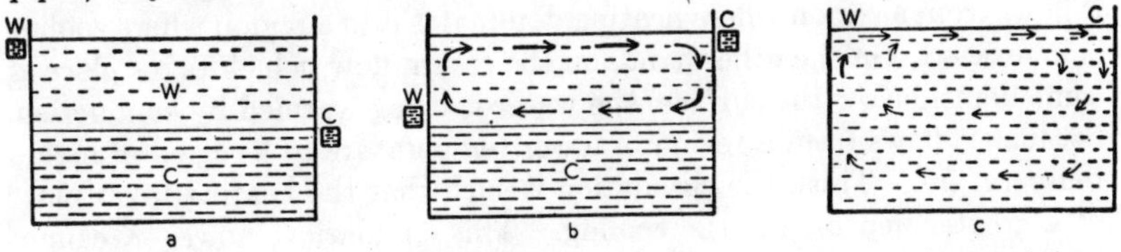

Fig. 39. Types of circulation induced in water by different placement of warm (W) and cold (C) sources.

until the water above the heater had been heated to the temperature of the circulating warm water and the water below the cooler had been cooled to the temperature of the circulating cold water. When this state had been reached and a stable stratification had been established, all motion ceased.

In a second experiment (fig. 39b), Sandström placed the cooling system above the heating system. In this case the final state showed a circulation with ascending motion above the heating unit and with descending motion below the cooling unit. Thus, a stationary circulation was developed because the heating took place at greater depth than the cooling.

From these experiments and from Bjerknes' theorem, it is immediately evident that conditions in the oceans are very unfavorable to the development of thermal circulations. Heating and cooling take place mainly at the same level—namely, at the sea surface, where heat is received by radiation from the sun during the day when the sun is high in the sky, or lost by long-wave radiation into space at night or when the sun is so low that the loss is greater than the gain, and where heat is received or lost by contact with air.

Because heating and cooling take place at the surface, one might expect that no thermal circulation could develop in the sea, but this is not true. Consider a vessel filled with water. Assume that heating at the surface takes place at the left-hand end and that, toward the right-hand end, the heating decreases, becoming zero at the middle of the vessel. Beyond the middle cooling takes place, reaching its maximum at the other end (fig. 39c). Under these conditions the heated water to the left will have a smaller density than the cooled water to the right, and will therefore spread to the right. Owing to the continuity of the system, water must rise near the left end of the vessel and sink near the right end, thus establishing a clockwise circulation that, at the surface, flows from the area where heating takes place to the area where cooling takes place. When stationary conditions have been established, the temperature of the water to the left must be somewhat higher than the temperature of the water to the right, owing to conduction from above.

This circulation is quite in agreement with Bjerknes' theorem, because at the surface the water that flows from left to right is being cooled, since it flows from a region where heating dominates into a region where cooling is in excess. On the other hand, on the return flow, which takes place at some depth below the surface, the water is being warmed by conduction, because it flows from a region of lower temperature to a region of higher temperature. Thus, the circulation is such that the heating takes place at a greater depth than the cooling. This circulation, however, cannot become very intensive, particularly because the heating within the return flow must take place by the slow process of conduction.

If the oceanic circulation is examined in detail, many instances are found in which the vertical circulation caused by the wind is such that the thermal machine runs in reverse, meaning that mechanical energy is transformed into heat, thus checking the further development of the wind circulation. When upwelling takes place, the surface flow will be directed from a region of low temperature to a region of high temperature, and the subsurface flow will be directed from high to low temperature. The thermal machine that is involved will consume energy and thus counteract a too-rapid wind circulation. Thus, in the Antarctic the thermal circulation will be directed at the surface from north to south and will counteract the wind circulation, which will be directed from south to north. On the other hand, systems are found within which the thermal effect tends to increase the wind effect and within which the increase of the circulation must be checked by dissipation of kinetic energy.

Thermohaline Circulation. So far, only thermal circulation has been considered, but it must be borne in mind that the density of the water depends on both its temperature and its salinity, and that in the surface layers the salinity is subject to changes caused by evaporation, condensation, precipitation, and addition of fresh water from rivers. In the open ocean the changes in density are determined by the excess or deficit of evaporation over precipitation. These changes in density may be in the same direction as those caused by heating and cooling, or they may be in opposite directions. When examining the circulation that arises because of the external factors that influence the density of the surface waters, one must take changes of both temperature and salinity into account and must consider not the thermal but the thermohaline circulation. For this reason, Bjerknes' theorem is better formulated as follows: *If a thermohaline circulation shall produce energy, the expansion must take place at a greater depth than the contraction.* In this form, the theorem can be used to determine if, within any given circulation, energy is gained or lost through thermohaline changes.

If thermal and haline circulations are separated, in some instances they work together and in others they counteract each other. The greatest heating takes place in the equatorial region, where, owing to excess precipitation, the density is also decreased by reduction of the salinity. In the latitudes of the subtropical anticyclones the heating is less, and, in addition, the density of the water is increased by excess evaporation. Between the Equator and the latitudes of the subtropical anticyclones, conditions are therefore favorable for the development of a strong thermohaline circulation. North and south of these latitudes the haline circulation will, however, counteract the thermal, because the density is decreased by excess precipitation but increased by cooling. There a weak thermohaline circulation might be expected.

In the absence of a wind system, one might expect a slow thermohaline circulation directed from the Equator to the poles at the surface and in the opposite direction at some subsurface depth. This circulation would be modified by the rotation of the earth and by the form of the ocean basins, but nothing can be said as to the character of the system of currents that would be developed under such conditions. It is probable, however, that the existing current system bears no similarity to the one that would result from such a thermohaline circulation, but is mainly dependent upon the character of the prevailing winds and the extent to which the circulation maintained by the wind is checked by the thermal conditions. In other words, the wind system tends to bring about a distribution of density that is inconsistent with the effect of heating and cooling, and the actual distribution approaches a balance between the two factors. These two factors—the wind and the process of heating and cooling—are variable, however, in time and space, and therefore a stationary distribution of density with accompanying stationary currents does not exist. Only when average conditions over a long time and a large area are considered can they be regarded as stationary.

VERTICAL CONVECTION CURRENTS. The thermohaline circulation is of small *direct* importance to the horizontal current, but is mainly responsible for the development of *vertical convection currents.* Wherever the density of the surface water is increased so much by cooling or by evaporation that it becomes greater than the density of the underlying strata, the surface water must sink and be replaced by water from some subsurface depth. The vertical currents that arise in this manner are called vertical convection currents. They are irregular in character and should not be called "currents" if this term is defined as motion of a considerable body of water in a definite direction.

The depth to which vertical convection currents penetrate depends upon the stratification of the water. A mass of surface water whose density has been increased by cooling or evaporation sinks until it meets water of the same density. If mixing with neighboring water masses takes place, it sinks to a lesser depth. When vertical convection currents have been active for some time, an upper layer of homogeneous water is formed, the thickness of which depends upon the original stratification of the water, the intensity of the convection currents, and the time the process has lasted. Thus, an upper homogeneous layer can be formed in two different manners—either by the mechanical stirring due to wind, or by the effect of thermohaline vertical convection currents.

The vertical convection currents as a rule are of greater importance in higher latitudes. In latitudes where an excess of evaporation is found, the heating of the surface is often so great that the decrease of the surface density by heating more than balances the increase by evaporation. In these circumstances the surface salinity will be greater than the salinity at a short distance below the surface. The formation of deep and bottom water by vertical convection currents is dealt with elsewhere (p. 89).

CHAPTER IX

Water Masses and Currents of the Oceans

Water Masses of the Oceans

In oceanography the introduction of the concept of water masses has been as fruitful as the introduction of the concept of air masses in meteorology. In the ocean, however, a steady state is more closely approached, and the distribution of the water masses deviates from the distribution of the air masses in that the boundaries between different water masses are always found in approximately the same localities and that the quasi-horizontal boundaries between water masses are as well defined and important as are the inclined boundaries.

In middle and low latitudes the arrangement of the water masses in a vertical direction is such that one can distinguish between the surface layer, the upper water, the intermediate water, the deep water, and, in some localities, the bottom water. In high latitudes the intermediate water is often lacking, and the upper water is similar to the deep water. In a vertical direction the density of the water increases with depth, and in a horizontal direction the density in general increases toward the polar regions. Surface water of a given density which sinks in higher or middle latitudes spreads along the proper density surface (σ_t surface); hence in middle latitudes the vertical distribution of density reflects the horizontal distribution during the season when the surface densities are greatest—that is, during the season when sinking of surface water is most likely to occur.

The following water masses are formed by *sinking* of surface water in different localities. *Antarctic Bottom Water* is formed near the Antarctic Continent, particularly in the Weddell Sea area to the south of the Atlantic Ocean. On the continental shelf the salinity of the water is increased by freezing of ice, so that water is formed at a salinity of about 34.62°/oo and a temperature of about $-1.9°$. Expressed as σ_t, the density of this water is 27.89, which is higher than the density of the adjacent circumpolar water, for which $\sigma_t = 27.84$, the water having a salinity of 34.68°/oo and a temperature of 0.5°. The water on the continental shelf therefore sinks, flowing down the continental slope, but in sinking it mixes with the warmer and more saline circumpolar water, so that a bottom water, the Antarctic Bottom Water, is formed with a salinity of about 34.66°/₀₀ and a temperature of about −0.4°. This water represents a *water type* (p. 88), but as it spreads it becomes mixed with adjacent water and attains the character of a *water mass*, which is marked by a certain temperature-salinity relation (p. 88).

North Atlantic Deep and Bottom Water is formed in the Labrador Sea and in the region between Iceland and the southern part of Greenland. There the warm and saline waters of the North Atlantic drift are mixed with the colder and less saline water of the East Greenland Current. In winter, when this mixed water is cooled at the surface, it attains a sufficiently high density (up to $\sigma_t = 27.88$) to sink. The water that sinks to depths greater than 1000 m has a salinity between 34.90 and 34.96°/₀₀ and a temperature between 2.8° and 3.3° (table 19, p. 170). Variations in the processes of mixing and cooling lead in different years to the sinking of water of somewhat different temperatures, salinities, and densities. The vertical convection currents therefore reach different depths and lead to the formation of a water mass within which the temperature decreases somewhat with depth. Corresponding deep water is not formed in the North Pacific, where no water of high salinity is subjected to intense cooling in winter.

Antarctic Intermediate Water sinks at the *Antarctic Convergence.* This convergence, the cause of which is not clearly understood, can be traced all around the Antarctic Continent within the belt of the strong and prevailing westerly winds. The water that sinks at the Convergence is a water type having everywhere a salinity of about 33.8°/₀₀ and a temperature of about 2.2°, the corresponding σ_t being 27.0. After the water leaves the surface, mixing with the over- and underlying bodies of water leads to the formation of a water mass that spreads to the north between the surfaces $\sigma_t = 27.2$ and $\sigma_t = 27.4$, and at its core is characterized by a salinity minimum.

North Atlantic Intermediate Water is formed in a similar manner at a convergence to the south of the Labrador Sea, but on a very small scale.

North Pacific Intermediate Water is similarly formed in the northeastern part of the North Pacific, in about latitude 40°N, but, when it spreads to the west and south, subsurface water is added in relatively large quantities. The North Pacific Intermediate Water is therefore lacking the characteristics that it would show if it were formed mainly by the sinking of surface water, such as a high content of dissolved oxygen. In contrast to the Antarctic and North Atlantic Intermediate Waters, it is poor in oxygen.

The *Central Water Masses* of the different oceans are formed by sinking at the Subtropical Convergences. The Subtropical Convergences are found in all oceans in latitudes 35° to 40° S and 35° to 40°N. They are not well-defined lines of convergence, but represent regions of convergence within which sinking of surface water to greater depths takes place in winter. In these regions the temperature and the salinity of the surface water decrease with increasing latitude, and the density increases. The sinking of surface water in these areas therefore leads to the formation of water masses whose temperature-salinity curves reflect the temperature-salinity relations at the surface in the coldest season, as shown schematically in fig. 21, p. 90. Because of the process of formation the character of the Central Water Masses in the different oceans depends upon the temperature and the salinity in the regions of the Subtropical Convergences.

In adjacent seas that are more or less closed off from the ocean by submarine ridges, special types of water are formed. Thus, in the Norwegian Sea, a deep water is formed by processes similar to those that lead to formation of the Atlantic Deep Water. In the western part of the Norwegian Sea, off the east coast of Greenland, a mixture of the warm Atlantic water that enters the Norwegian Sea to the north of Scotland and the cold water of the East Greenland Current is cooled in winter, and, by vertical convection currents, deep water is formed of temperature $-0.8°$ to $-1.2°$ and of salinity 34.89 to 34.92°/₀₀. The upper layers of this water flow into the Polar Sea across the submarine ridge between Greenland and Spitsbergen and fill the Polar Basin.

In the Mediterranean Sea, evaporation and winter cooling lead to the formation of a deep water of temperature 13.0° to 13.6° and of salinity 38.4°/₀₀ to 38.7°/₀₀. This *Mediterranean Water* flows along the bottom of the Strait of Gibraltar into the Atlantic Ocean, where it mixes with the Atlantic water and spreads between the surfaces $\sigma_t = 27.5$ and $\sigma_t = 27.7$—that is, below the Antarctic Intermediate Water. It can be clearly recognized over wide areas by an intermediate salinity maximum; in the South Atlantic, it can be traced around the Cape of Good Hope.

In the Red Sea a similar but warmer and more saline deep water is formed, having a temperature of 21.5° to 22° and a salinity of 40.5°/₀₀ to 41°/₀₀. This *Red Sea Water* flows along the bottom of the Strait of Bab-el-Mandeb into the Indian Ocean, where it mixes with other water masses and spreads. The quantities of Red Sea Water entering the Indian Ocean are much smaller than the quantities of Mediterranean Water entering the Atlantic Ocean, and therefore the Red Sea Water exercises a smaller influence.

Other important water masses in the oceans are formed not by sinking of surface water but by processes of *subsurface mixing*. The greatest of all is the *Antarctic Circumpolar Water Mass*, which is mainly formed by mixing of Atlantic deep water and Antarctic bottom water, but also contains some Antarctic Intermediate Water and traces of Mediterranean Water. The processes that lead to the formation of this water mass are discussed on pp. 214 and 215.

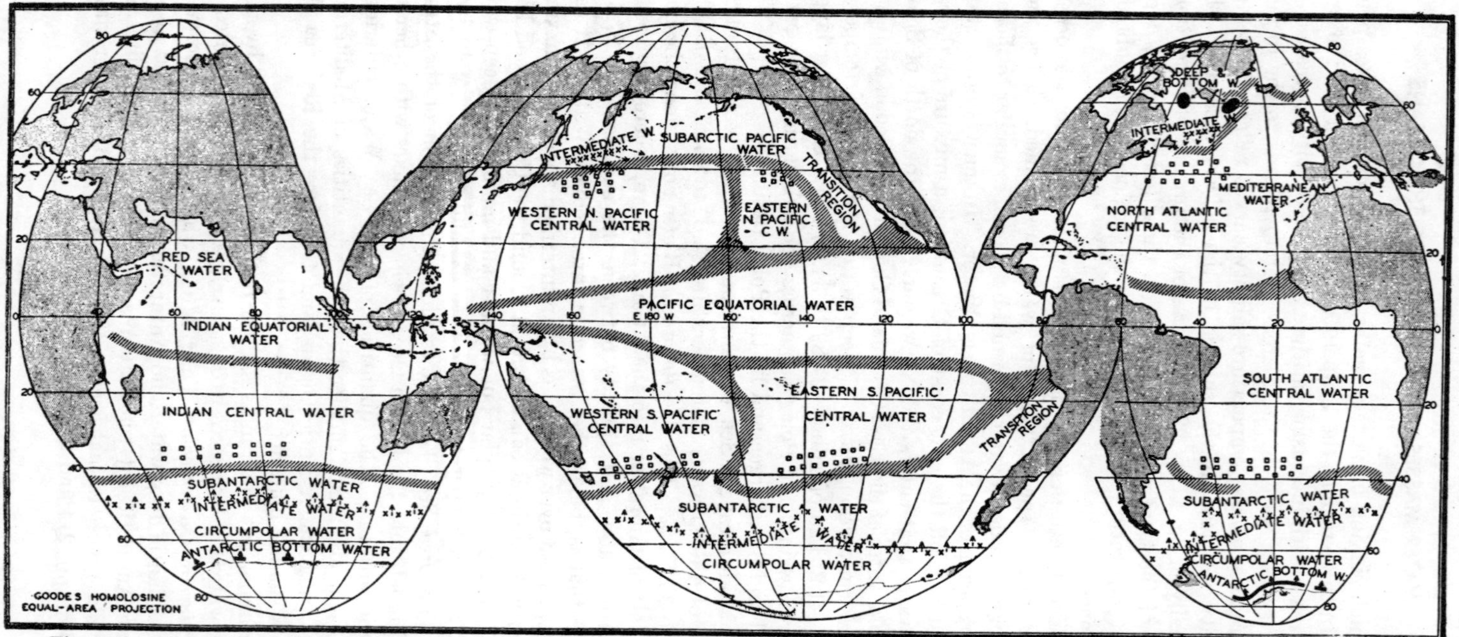

Fig. 40A. Approximate boundaries of the Upper Water Masses of the ocean. Squares indicate the regions in which the Central Water Masses are formed, and crosses show the lines along which the Antarctic and Arctic Intermediate Waters sink.

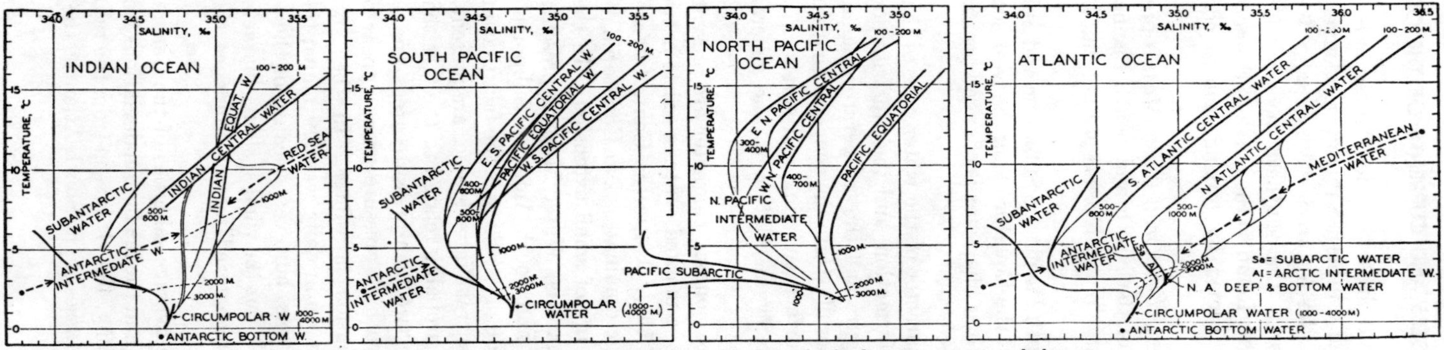

Fig. 40B. Temperature-salinity relations of the principal water masses of the oceans.

The *Subantarctic Water Mass* occupies the region between the Antarctic Convergence and the southern Subtropical Convergence, and represents a transition from the Antarctic Circumpolar Water Mass to the Central Water Masses of the southern oceans. The corresponding *Subarctic Water Masses* are found to the north of the northern Subtropical Convergences and are formed by mixing, winter cooling, and excess precipitation.

In the Pacific Ocean an *Equatorial Water Mass* is present between the Central Water Masses of the North and South Pacific. In the Indian Ocean a similar Equatorial Water Mass is present, but it is lacking in the North Atlantic.

In the North and South Pacific the Subarctic and Subantarctic Water Masses penetrate toward the Equator along the west coasts of North and South America, where their character is changed because of heating and evaporation at the surface.

Fig. 40 shows the character of the water masses that have been discussed, their regions of formation, and their distribution. The chart in the upper part of the figure and the temperature-salinity curves in the lower part should together illustrate the concepts that most water masses are formed at the sea surface and sink and spread in a manner which depends upon their density in relation to the general distribution of density in the oceans.

Water masses of corresponding character are present in the different oceans, but, nevertheless, a marked contrast exists between the Atlantic and Pacific Oceans. In the Atlantic the Central Water Masses of the North and South Atlantic dominate. No Equatorial Water exists, and Subpolar Water Masses are of small extension. In the Pacific, on the other hand, a large Equatorial Water Mass is present, and, along the western coasts of North and South America, Subpolar Waters advance toward the Equator. Equally striking is the fact that in both the North and South Pacific two Central Water Masses are present. The existence of these water masses is probably related to the fact that over both the North and South Pacific two high-pressure areas are generally present. In the South Pacific, two high-pressure areas appear on the charts of average pressure distribution in winter. In the North Pacific, two pressure areas often show up in daily weather maps; these areas do not appear on charts for seasons, however, because of the variable location of the eastern pressure cell, but the wind systems characteristic of the two high-pressure areas are recognized by the influence they exert on the oceanographic conditions.

Table 17 contains the average salinities of the Central Water Masses at different temperatures and the maximum deviations from the averages, according to the curves in fig. 40. It is seen from the table and the figure that the Central Water Masses of the South Atlantic, the Indian, and the

TABLE 17

SALINITIES OF THE CENTRAL WATER MASSES OF THE OCEANS IN PARTS PER THOUSAND AT STATED TEMPERATURES

Temperature (°C)	8°	10°	12°	14°	16°
South Atlantic	34.64 ± 0.08	34.86 ± 0.08	35.11 ± 0.08	35.37 ± 0.09	35.64 ± 0.10
Indian Ocean	34.65 ± 0.07	34.89 ± 0.06	35.13 ± 0.07	35.37 ± 0.08	35.62 ± 0.09
Western South Pacific	34.58 ± 0.07	34.80 ± 0.06	35.04 ± 0.08	35.30 ± 0.08	35.55 ± 0.09
Eastern South Pacific		34.54 ± 0.07	34.70 ± 0.08	34.88 ± 0.08	35.08 ± 0.08
North Atlantic	35.12 ± 0.09	35.37 ± 0.09	35.63 ± 0.09	35.88 ± 0.09	36.12 ± 0.09
Western North Pacific		34.24 ± 0.07	34.38 ± 0.06	34.52 ± 0.06	34.67 ± 0.07
Eastern North Pacific			34.11 ± 0.09	34.32 ± 0.08	34.62 ± 0.08

western South Pacific Oceans are very similar, as should be expected, because they are formed in regions in which the external influences—that is, the atmospheric circulation and the processes of heating and cooling—are similar. The corresponding water mass of the eastern South Pacific is of lower salinity, probably because of admixture of the low-salinity Subantarctic Water of the Peru Current. Such admixture may also be responsible for the fact that the Central Water of the western South Pacific has a slightly lower salinity than the Central Waters of the Indian and South Atlantic Oceans.

The Central Waters of the North Atlantic and the North Pacific Oceans are quite different, the former having a very high and the latter a very low salinity. The contrast probably results from the different character of the ocean circulation and the differences in the amounts of evaporation and precipitation, especially in high latitudes, which are intimately related to the distribution of land and sea.

The Central Water Masses are all of small vertical extension, particularly in the North Pacific Ocean, where their thickness over large areas is only 200 to 300 m. In all oceans the greatest thickness of the Central Water Masses is found along the western boundaries, reaching 900 m in the Sargasso Sea region of the North Atlantic.

The Central and Equatorial Water Masses are covered by a surface layer 100 to 200 m thick, within which the temperature and the salinity of the water vary greatly from one locality to another, depending upon the character of the currents and the exchange with the atmosphere, and within which great seasonal variations occur in middle latitudes. The surface layer, the Central Water Masses, and the upper portions of the Equatorial Water Masses together form the oceanic troposphere (p. 86).

The Subantarctic Water between the Central Water Masses of the southern oceans and the Antarctic Convergence has nearly the same character all around the earth, and is therefore considered as belonging to the waters of the Antarctic Ocean. It is of low salinity and is probably formed by mixing and vertical circulation in the region between the Subtropical and the Antarctic Convergences. In the North Atlantic the corresponding Subarctic Water is found only in a small region and is of relatively high salinity, but in the North Pacific it is of wide extension and of low salinity. The Subarctic Water must be formed by processes that differ from those that maintain the Subantarctic Water. In the southern oceans the Antarctic Convergence represents a continuous and well-defined poleward boundary of the Subantarctic Water, but in the northern oceans the corresponding Arctic Convergence is found in the western parts of the oceans only, and in large areas there is no marked poleward boundary of Subarctic Waters. This contrast between south and north must be related to the differences in the distribution of land and sea and is reflected in the character of the waters. The Subarctic Waters are similar to the corresponding Arctic Intermediate Waters, but the Subantarctic Water is distinctly different in character from the Antarctic Intermediate Water.

The regions of formation of the intermediate, deep, and bottom waters are shown in the chart in fig. 40, and the characteristic temperature and salinity values of these water masses can be read off from the diagrams in the lower part of the figure. The deep water of the different oceans will be discussed more fully when dealing with the deep-water circulation of the oceans.

Currents of the North Atlantic Ocean

THE NORTH EQUATORIAL CURRENT. The system of currents in the North Atlantic (chart 4) is dominated by the North Equatorial Current to the south and by the Gulf Stream system to the north. The North Equatorial Current flows from east to west in the trade-wind region and is fed by the southeasterly currents off the west coast of North Africa. Corresponding to flow from the northwest, water of relatively high density and low temperature is found off the African coast, as is evident from the charts of surface temperatures (charts 1 and 2). The temperature close to the coast is also lowered by upwelling from moderate depths, owing to the action of prevailing northwesterly winds, but this upwelling does not exercise as wide-spread an influence as does the corresponding upwelling off the coasts of southwest Africa, or, particularly, as that off the west coasts of North and South America.

In the western part of the North Atlantic Ocean the North Equatorial Current joins a branch of the South Equatorial Current which has crossed the Equator and which, according to fig. 40, p. 159, carries characteristically different water masses. Thus, the part of the North Equatorial Current that continues into the Caribbean Sea carries water which is mixed with water of South Atlantic origin, whereas the northern branch of the North Equatorial Current, which flows along the northern side of the Great Antilles as the Antilles Current, carries water that is identical with that of the Sargasso Sea.

THE GULF STREAM SYSTEM. The North Equatorial Current terminates in the current through the Yucatan Channel and the Antilles Current, and the continuation of these currents represents the beginning of the Gulf Stream system, which dominates the circulation of a great part of the North Atlantic Ocean. In accordance with the nomenclature of Iselin the term "Gulf Stream system" is used to include the whole northward and eastward flow beginning at the Straits of Florida and including the various branches and whirls that are found in the eastern North Atlantic and that can be traced back to the region south of the Newfoundland Banks. This system can be subdivided into the following parts:

(1) The Florida Current, comprising all the northward-moving water from the Straits of Florida to a point off Cape Hatteras, where the current ceases to follow the continental slope. The Florida Current can be traced directly back to the Yucatan Channel, because the greater part of the water flowing through this strait continues on the shortest route to the Straights of Florida and only a small amount sweeps into the Gulf of Mexico, later to join the Florida Current. After passing the Straits of Florida the current is reinforced by the Antilles Current, but the name "Florida Current" is retained as far as to Cape Hatteras.

(2) The Gulf Stream, comprising the mid-section of the system from the region where the current first leaves the continental slope off Cape Hatteras to the region to the east of the Grand Banks in about longitude 45°W, where the stream begins to fork. This application of the name "Gulf Stream" represents a restriction of the popular term, but such a restriction is necessary in order to introduce clear definitions.

(3) The North Atlantic Current, comprising all the easterly and northerly currents of the North Atlantic from the region to the east of the Grand Banks, where the Gulf Stream divides. The branches of the North Atlantic Current are often masked by shallow and variable wind-drift surface movements that have become commonly known as the North Atlantic Drift.

The terminal branches of the Gulf Stream system are not all well known, but among the major ones are the Irminger Current, which flows toward the west off the south coast of Iceland, and the Norwegian Current, which enters the Norwegian Sea across the Wyville Thomson Ridge and which ultimately can be traced into the Polar Sea. Other more irregular branches turn to the south and terminate in great whirls off the European coast.

THE FLORIDA CURRENT. The energy of the Florida Current appears to be derived directly from the difference in sea level between the Gulf of Mexico and the adjacent Atlantic coast, the observed difference between Cedar Keys and St. Augustine being 19 cm. Assuming that this hydrostatic head accounts for all of the energy, and assuming frictionless flow, Montgomery finds that the velocity through the Straits of Florida should be 193 cm/sec, which is somewhat higher than the average velocity at the center of the current. The difference in level is probably maintained by the trade winds, and *the energy of the Florida Current is therefore derived from the circulation of the atmosphere.*

Within the current flowing through the Straits of Florida the distribution of density must adjust itself in the usual manner so that the lighter water will be found on the right-hand side of the current and the denser water on the left-hand side, and so that the sea surface, instead of coinciding with a level surface, rises toward the right-hand side of the current. Across the Florida Current, this rise amounts to about 45 cm, so that at the coast of Cuba the sea level is about 45 cm higher than at the American mainland.

The character of the currents in the Straits of Florida was established by the outstanding measurements which, in the years 1885 to 1889, were made by the U. S. Coast and Geodetic Survey from the survey vessel *Blake*, commanded by J. E. Pillsbury. Pillsbury's current observations, which were carried out from a vessel anchored in deep water in a swift stream, are among the classical data in physical oceanography, not so much because they give complete information as to the average currents, but mainly because they made possible a convincing demonstration of the correctness of the later methods used for computing relative currents. The upper right-hand graph in fig. 26 (p. 111) shows the observed average velocity distribution in a section through the narrowest part of the Straits of Florida between Fowey Rocks, south of Miami, Florida, and Gun Cay, south of Bimini Islands, as plotted by Wüst from Pillsbury's data. To the left are shown the corresponding distributions of temperature and salinity as represented by Wüst on the basis of temperature measurements made by Bartlett on board the *Blake* in 1878 and published by Agassiz in 1888, and of temperature and salinity observations on board the U. S. Coast and Geodetic Survey vessel *Bache* in 1914. By means of these data, Wüst has computed the velocity distribution which is shown in the lower right-hand graph in fig. 26 and which is in remarkable agreement with the observed distribution. In order to arrive at absolute values of the velocity, Wüst had to assume a known velocity at some depth, and, on the basis of the distribution of temperature and salinity, he assumed an inclined surface of no motion at some distance from the bottom, as shown by the curves marked *O*. These curves nearly coincide with the curve of zero velocity as derived from Pillsbury's measurements. Thus, a complete correspondence is found between observed and computed currents, and this single example, therefore, has greatly contributed to increasing the confidence in the correctness of computed currents in general.

On the basis of measurements and computations, Wüst finds that the average transport of water through the Straits of Florida is 26 million m^3/sec. The transport probably shows an annual variation and may differ from year to year, but so far little is known about such fluctuations.

In its further course the Florida Current closely follows the continental slope, flowing most swiftly directly along the slope. The shallow coastal waters to the left of the Florida Current remain more or less at rest, and often the transition from these waters to the blue waters of the Florida Current is so abrupt that the border of the Florida Current can be seen as a line stretching from horizon to horizon. After emerging from the Straits of Florida the current is soon joined by the Antilles Current, which, according to Wüst, carries about 12 million m^3/sec. Owing to the moderate depth to the bottom, the current remains relatively shallow, not more than about 800 m deep, and carries no water colder than 6.5° until it leaves the Blake Plateau in about latitude 33°N. According to Iselin the current increases steadily in volume by absorption of Sargasso Sea water, and as it leaves the Blake Plateau the depth and the volume of the current suddenly increase by the joining in of water of a temperature considerably below 8° that comes from the southwestern Sargasso Sea.

THE GULF STREAM. The middle portion of the Gulf Stream system, for which the name "Gulf Stream" is retained, continues as a well-defined and relatively narrow current which, in contrast to the Florida Current, flows at some distance beyond the continental shelf. To the right of the current is the Sargasso Sea water, as previously, but to the left are now found two water types: the coastal water, which covers the shallow shelf areas, and the slope water, which, at temperatures between 4° and 10°, is very similar to the Gulf Stream water, but which at higher temperatures is of lower salinity. Within the upper layers of the slope water, great seasonal variations in temperature and salinity occur, and, in addition, eddies of Gulf Stream water occasionally intrude. The surface velocities are very high, the computed values reaching, in lat. 36°N, long. 73°W, more than 120 cm/sec, and in lat. 38°N, long. 69°W, reaching 140 cm/sec. On the assumption of no motion at a depth of 2000 m, where the isosteres are nearly horizontal, the volume transport of the Gulf Stream off Chesapeake Bay is between 74 and 93 million m^3/sec. If these figures are correct, they indicate that between 38 and 57 million m^3/sec of Sargasso Sea water and deep water have been added to the Florida-Gulf Stream after the Antilles Current, carrying 12 million m^3/sec, joined the flow of 26 million m^3/sec through the Straits of Florida. Similarly, between 34 and 53 million m^3/sec would have to be discharged toward the south from the Gulf Stream between Chesapeake Bay and longitude 45°W, off the "tail" of the Grand Banks, where, according to Soule, the transport of the Gulf Stream is somewhat less than 40 million m^3/sec. These conclusions are not supported by observations between the line Chesapeake Bay-Bermuda and the Bahamas or between Bermuda and longitude 45°W. The available data indicate that the inflow north of the Antilles Current does not exceed 15 to 20 million m^3/sec, and between Bermuda and longitude 45°W the southward flow of Gulf Stream water does not exceed 15 million m/sec. The computed transport can, however, be interpreted differently.

The dynamics of the Florida Current and the Gulf Stream, particularly the downstream increase in volume as far as Cape Hatteras, is not clearly understood. Rossby has compared the Florida Current and its continuation, the Gulf Stream, to a wake stream which emerges from the Straits of Florida, and has examined the effect on such a stream of stresses due to lateral mixing. In a wake stream in homogeneous water the momentum transport remains constant, whereas the volume (mass) transport increases downstream, the increase being due to inflow from the sides. Expanding the theory to a stratified medium, Rossby finds that a "compensation current" in the direction of flow must develop on the right-hand side of the wake stream, whereas to the left a counter-current in the opposite direction must appear, and this picture agrees in general with the pattern of the Gulf Stream and its surroundings. Another important aspect of the theory is that, owing to the lateral stresses, a transverse circulation should develop, water being absorbed from the oceanic areas to the right of the current and discharged into the counter-current to the left. Such a mechanism would account for the presence of eddies of Gulf Stream water in the slope current, but Defant and Ekman have warned against immediate acceptance of the theory, because several of the necessary assumptions appear not to be fulfilled. Regardless of whether the theory is confirmed or disproved, it has been greatly stimulating, particularly because of its emphasis on the importance of lateral mixing.

TABLE 18

AVERAGE SEA LEVEL ALONG THE NORTH AMERICAN EAST COAST REFERRED TO SEA LEVEL AT THE COAST OF FLORIDA-GEORGIA

Locality	Sea level (cm)		Mean	Distance along coast (km)	Slope
	Avers, 1927	Rappleye, 1932			
St. Augustine, Fla.; Fernandina, Fla.; Brunswick, Ga.	0	0	0	0	
					6×10^{-8}
Norfolk, Va.	4	7	6	1000	
					35×10^{-8}
Cape May, N.J.; Atlantic City, N.J.; Fort Hamilton, N.J.	16	24	20	1400	
					13×10^{-8}
Boston, Mass.; Portland, Me.	25	30	28	2000	
					12×10^{-8}
Halifax, Nova Scotia	—	35	35	2600	

A satisfactory theory of the Gulf Stream must account not only for the increase in volume transport in the direction of flow and the fact that this increase takes place without evidence of strong inflow from the southeast, but also for another important feature that has been given considerable attention without having been explained satisfactorily. Precise leveling along the American east coast shows that the mean sea level increases toward the north from St. Augustine, Florida, to Halifax, Nova Scotia, the most conspicuous increase taking place directly north of Cape Hatteras. In table 18, summarizing the results of precise leveling, the sea level along the coast has been referred to that on the coast of Florida and southern Georgia, values from stations less than 200 km apart having been combined as averages. The distances along the coast from Florida-Georgia are entered, and also the values of the slope of the sea surface. These values are of the same order of magnitude as those derived from oceanographic data in the Caribbean Sea, where Parr computed a slope of 17×10^{-8}, and Sverdrup a slope of 12×10^{-8} (p. 181).

If the upward slope along the American east coast were due to the distribution of mass in the ocean, the average density of the Gulf Stream water off the continental shelf would have to decrease in the direction of flow and the Gulf Stream would have to flow uphill, but no such decrease is found. A discrepancy therefore exists between the results of precise leveling and the results of what has been called "oceanographic leveling." It is not surprising that such discrepancies appear because, as explained on p. 99, oceanographic observations can give information only as to the topography of the sea surface *relative* to some selected surface in the ocean, and information as to the *absolute* topography of the sea surface must be derived from precise leveling along the coasts. It seems possible, however, to reconcile the different observations by assuming that an actual piling up of water is maintained in the western North Atlantic and that, owing to this piling up, a current flowing southwest along the continental shelf is not reflected in the distribution of mass.

THE NORTH ATLANTIC CURRENT. The North Atlantic Current represents the continuation of the Gulf Stream after it leaves the region to the east of the "tail" of the Grand Banks. Beyond this region the Gulf Stream loses its characteristics as a well-defined current and divides into branches that are often separated by countercurrents or eddies. Some of the branches turn south, but others continue toward the east across the mid-Atlantic Ridge, being flanked on the northern side by waters of the Labrador Current that have been mixed with Gulf Stream water.

The contrast between the Gulf Stream and the North Atlantic Current to the north of the Azores is illustrated in fig. 41, which shows two temperature profiles on the same scale. To the left in the figure are shown the isotherms in a vertical section from Chesapeake Bay toward Bermuda, according to observations at *Atlantis* stations 1231–1226. The Gulf Stream is here concentrated within the narrow band in which the isotherms slope steeply downward toward the right. The section to the right in the figure runs north-northwest from the Azores to lat. 48°N and is based on the observations on board the *Altair* during the International Gulf Stream Expedition in 1938. In this section the isotherms generally slope downward toward the south, indicating a flow toward the east, but the slope is not uniform and countercurrents or eddies are present between the east-flowing branches of the current.

The detailed work that was conducted by the International Gulf Stream Expedition of 1938 clearly shows the complicated details of the oceanographic conditions. Between June 1 and 22, 1938, the German vessel, the *Altair*, and the Norwegian vessel, *Armauer Hansen*, occupied 159 stations in an area of less than 100,000 square miles, and at one station the *Altair* anchored and made hourly observations of temperature, salinity, and currents at a number of depths between the surface and 800 m for a period of 90 hours. The dense network of stations showed

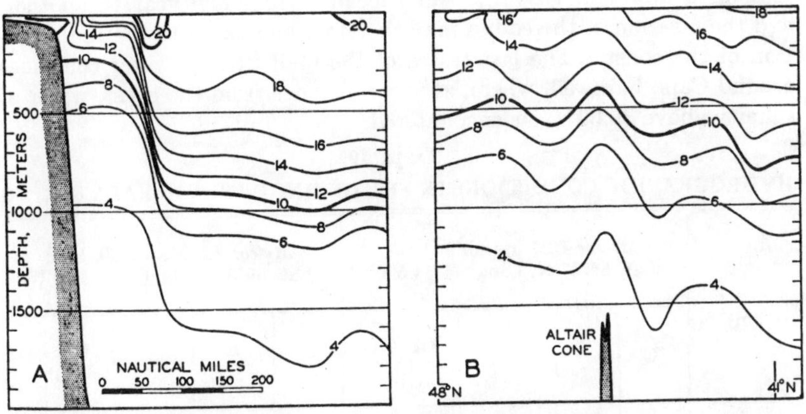

Fig. 41. Temperature profiles across the Gulf Stream off Chesapeake Bay and across the North Atlantic Current to the north of the Azores.

even greater irregularities than one might expect. Thus, at 600 m, the temperature varied between approximately 7° and 13°, and differences up to 5° were observed at distances of less than 40 miles. Some of the observed features may be due to the influence of the bottom topography, and others may be related to traveling disturbances. Regardless of how the features are interpreted, they do show that caution has to be exercised when drawing conclusions from a few scattered observations, because such scattered data may not be representative. They also show that an intensive mixing takes place in mid-ocean.

Records at stations that were occupied in 1931 by the *Atlantis* along the mid-Atlantic Ridge in long. 30°W demonstrate that in spite of irregularities one can distinguish between two major branches of the North Atlantic Current. The northern of these branches flows between latitudes 50° and 52°N, at the boundary between the water of the Gulf Stream system and the Subarctic Water, and carries water which represents Gulf Stream water mixed with waters of the Labrador Current.

The other branch flows approximately in lat. 45°N and carries undiluted Gulf Stream water. Of these branches the northern continues mainly toward the east-northeast and divides up. Part of the water flows across the Wyville Thomson Ridge into the Norwegian Sea, and part turns toward the north and northwest as the Irminger Current, which flows along the southern coast of Iceland. A small portion of the water of the Irminger Current bends around the west coast of Iceland (fig. 44, p. 175), but the greater amount turns south and becomes more or less mixed with the waters of the East Greenland Current. The detailed work of the *Meteor* in the area between Iceland and southeastern Greenland indicates that many eddies within which this mixing takes place remain in approximately the same locality from one year to another, and the position of the eddies may therefore be related to the configuration of the coast. The last traces of the Gulf Stream water continues around Cape Farewell, where, at some distance from the coast, water of salinity above 35.00°/₀₀ is encountered.

TABLE 19

HYDROGRAPHIC CONDITIONS IN THE IRMINGER SEA IN EARY SPRING

Depth (m)	*Meteor* 121, March 9, 1935 Lat. 56°37′N, Long. 44°54.5′W					*Meteor* 79, March 30, 1933 Lat. 59°38′N, Long. 40°42.5′W				
	°C	S (°/₀₀)	σ_t	O_2 (ml/L)	O_2 (%)	°C	S (°/₀₀)	σ_t	O_2 (ml/L)	O_2 (%)
0......	2.82	34.87	27.80	7.24	96	4.07	34.96	27.76	6.92	95
50......	3.01	34.90	27.81	7.26	97	4.07	34.97	27.77	6.99	96
100......	3.09	34.92	27.82	7.24	97	4.06	34.96	27.77	6.81	94
200......	3.17	34.93	27.82	6.70	90	3.98	34.97	27.77	6.84	94
800......	3.26	34.96	27.83	6.98	94	3.76	34.95	27.77	6.60	90
1000......	3.20	34.95	27.83	6.96	93	3.33	34.89	27.77	6.64	89
1500......	3.17	34.93	27.82	6.99	94	3.28	34.94	27.82	6.39	86
2000......	3.23	34.93	27.82	—	—	2.84	34.96	27.88	6.37	87

In the central part of the Irminger Sea, off southern Greenland, mixing between the North Atlantic Water and the Labrador Sea Water leads to the formation of Subarctic Water of uniform salinity close to 34.95°/₀₀ and a temperature which, from a depth of a few hundred meters and downward, is nearly 3°C. When the surface layers are cooled in winter to temperatures below 3°, vertical convection currents develop, reaching from the surface to the bottom and leading to renewal of the deep water. Table 19 contains data from two *Meteor* stations which were occupied in early spring and at which nearly uniform water was present. In the table the oxygen content of the water is stated and is expressed both as milliliter of oxygen per liter of water and as percentage of saturation at atmospheric pressure. The high oxygen values substantiate the conclusion that convection currents were present and reached to great depths.

The branch of the North Atlantic Current which crosses the mid-Atlantic Ridge in approximately lat. 45°N turns to the right and continues as an irregular flow toward the south between the Azores and Spain, sending whirls into the Bay of Biscay. No distinct currents exist in this region, as is evident from the discussions of the conditions in the eastern North Atlantic by Helland-Hansen and Nansen, but a diffuse transport of

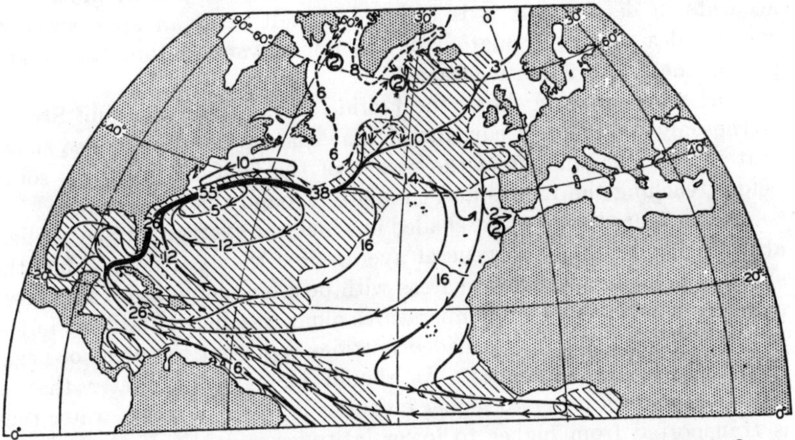

Fig. 42. Transport of Central Water and Subarctic Water in the Atlantic Ocean. The lines with arrows indicate the direction of the transport, and the inserted numbers indicate the transported volumes in millions of cubic meters per second. Full-drawn lines show warm currents, dashed lines show cold currents. Areas of positive temperature anomaly are shaded.

water toward the south takes place. Some of this water enters the Mediterranean as a surface current and flows out again across the sill in the Strait of Gibraltar as water of very high salinity which spreads at intermediate depths and there influences conditions over the greater part of the North and South Atlantic Oceans (p. 173). The larger amount of the upper water masses continues toward the south and finally joins the North Equatorial Current.

Transport. On the basis of the above discussion and from numerous computations of transports, fig. 42 has been prepared giving a schematic picture of the volume transport of the currents that carry North Atlantic Central Water or Subarctic Water. No lines of equal transport are entered, but the different branches of the current system are indicated and the approximate volume transports in millions of cubic meters per second are stated. The presentation has been derived by computing the transport of water of temperature higher than 7° between a number of selected stations north of lat. 20°N, and adjusting the figures in order to take the continuity of the system into account. To the northwest of a line that can be drawn roughly between the Straits of Florida and the English Channel it has been assumed that the 2000-decibar surface could be taken as a surface of no motion, but to the southeast of that line it has been assumed in general that the 7°-isothermal surface was a surface of no motion. This treatment was necessary because the condition of continuity has to be satisfied and because a reversal of the direction of flow appears to take place below the 7°-isothermal surface over large parts of the southeastern area. It is not claimed that the picture in fig. 42 is accurate in details, but it is believed that it gives an approximately correct idea of the circulation of the upper water within the greater part of the North Atlantic Ocean.

The representation shows the Florida Current and the Gulf Stream as the only well-defined currents of the Atlantic Ocean. It also shows that the greater amount of the waters of the Gulf Stream turns south before reaching the Azores and circulates around the Sargasso Sea.

In the figure the areas are shaded in which the average surface temperature is higher than the general average for that latitude. Thus the shaded portions represent the areas with positive temperature anomalies as referred to the average temperatures along parallels of latitude in the Atlantic Ocean, and the unshaded portions represent the areas of negative temperature anomalies. As should be expected, water that is transported from lower to higher latitudes is warm, whereas water that is transported from higher to lower latitudes is cold. If the velocity distribution within the different branches of the currents were known, it would be possible to compute the net amounts of heat that are given off or taken up by the ocean currents, but so far such calculations must be made on an entirely different basis (p. 229).

Another characteristic of the system of currents is brought out by a comparison between the transports and the temperatures off Chesapeake Bay and to the north of the Azores (fig. 41). The Gulf Stream off Chesapeake Bay transports large volumes of water of temperatures above 16°, but the North Atlantic Current to the north of the Azores carries only small amounts of water as warm as that. This apparent reduction in temperature must be due to the fact that the warmer waters of the upper layers have been carried south before reaching the Azores, whereas the somewhat colder waters at greater depths have risen and continued toward the east.

As will be shown (p. 186), 6 million m^3/sec of South Atlantic Upper Water enter the North Atlantic Ocean along the coast of South America. Correspondingly, 6 million m^3/sec of North Atlantic Water sink in different localities and return to the South Atlantic as a deep-water flow. The three regions where sinking takes place and the amounts of water sinking from the surface are shown by circles and inserted numbers that represent rounded-off values. About 2 million m^3/sec sink outside the Strait of Gibraltar and about the same amount sinks in the Labrador Sea. Both of these values are based on fairly accurate data, wherefore it follows that a similar amount sinks in the third region within which bottom and deep water is being renewed—namely, the region to the southeast of southern Greenland.

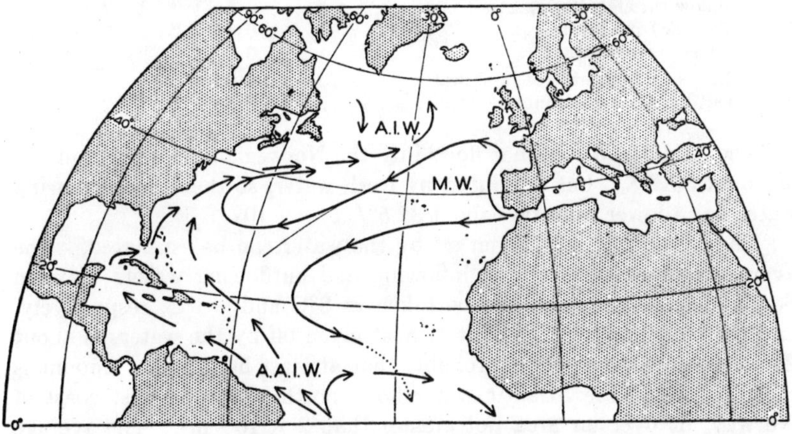

Fig. 43. Approximate directions of flow of the Intermediate Water Masses of the North Atlantic. A.I.W.: Arctic Intermediate Water, M.W.: Mediterrenean Water, A.A.I.W.: Antarctic Intermediate Water.

The direction of flow of the different types of intermediate waters of lower temperature does not always coincide with the direction of flow of the upper water. On the basis of the character of the water and the results of dynamic calculations, fig. 43 has been prepared giving a tentative picture of the spreading of the Arctic Intermediate Water, the Mediterranean Water, and the Antarctic Intermediate Water. The Arctic Intermediate Water is present in the northern region only. The Mediterranean Water partly bends north before turning west and partly spreads directly toward the west. Some of this water turns south and continues across the Equator below the Antarctic Intermediate Water, as indicated by the crossing of the lines of flow. The Antarctic Intermediate Water enters the North Atlantic Ocean along the coast of South America. One branch bends toward the east and south, returning across the Equator, and two other branches continue, one into the Caribbean Sea, and the other along the north side of the Antilles. Mixing takes place between the different types of water, and the actual pattern of flow or spreading out is therefore far more complicated than shown.

Currents of the Adjacent Seas of the North Atlantic Ocean

THE NORWEGIAN SEA AND THE NORTH POLAR SEA. The inflow of water into the Norwegian Sea and the North Polar Sea takes place principally between Scotland and the Faroe Islands, and the outflow takes place through Denmark Strait. Fresh water is added by run-off from the great Siberian rivers and by excess precipitation. The water budget of the region can therefore be summarized as follows:

Inflow northwest of Scotland	3.0	million m^3/sec
Inflow through Bering Strait	0.3	" "
Run-off from rivers	0.16	" "
Excess precipitation	0.09	" "
Inflow and addition of fresh water	3.55	" "
Outflow through Denmark Strait	3.55	" "

The Atlantic Water that flows into the Norwegian Sea has a salinity of about 35.3°/₀₀, but is diluted by fresh water, so that the outflowing water has a lower salinity—about 32.5°/₀₀.

The amount of heat given off by the water can be estimated. The average temperatures of the inflowing and outflowing waters between Scotland and Greenland can be taken as 8°C and −1°C, respectively, and therefore the total amount of heat given off by the waters is about 24×10^{12} g cal/sec. It is probable that at least half of this amount is given off where the Atlantic water flows north along the west coast of Norway, or over an area not greater than 2×10^{12} m^2. The average amount of heat given off in this area would then be 12 g cal/m^2/sec = 0.072 g cal/cm^2/min = 103 g cal/cm^2/day. This value is greater than the radiation surplus in those latitudes (fig. 66 p. 229), and, although it is uncertain, it serves to demonstrate the important bearing of the Atlantic Current on the climate of the extreme northwestern part of Europe.

Later observations have not materially changed the conception of the surface currents of the Norwegian Sea which Helland-Hansen and Nansen presented in 1909. Their picture is reproduced in fig. 44, according to which the two main currents in that region are the Norwegian Current, which represents a continuation of the North Atlantic Current, and the East Greenland Current.

The Norwegian Current, which is part of the Gulf Stream system (p. 164), is flanked on the left-hand side by a series of whirls, some of which are probably stationary and related to the bottom topography, whereas others may be traveling eddies. Off northern Norway the Norwegian Current branches, one branch continuing into Barents Sea, and another turning north toward Spitsbergen and bending around the northwestern Spitsbergen islands. The waters of this current have a high salinity and a high temperature, the maximum salinity decreasing from about 35.3°/₀₀ north of Scotland to about 35.0°/₀₀ off Spitsbergen and the subsurface temperature decreasing from about 8° to less than 4° in the same distance.

The East Greenland Current flows on or directly off the East Greenland shelf, carrying water of low salinity and low temperature. The

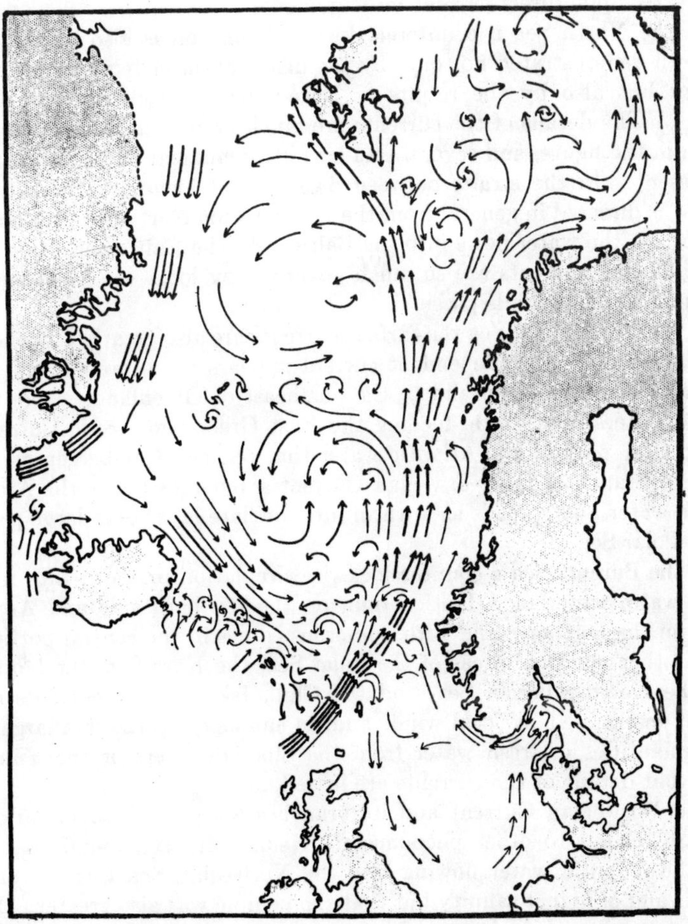

Fig. 44. Surface currents of the Norwegian Sea (after Helland-Hansen and Nansen).

greater part of the East Greenland Current continues through Denmark Strait between Iceland and Greenland, but one branch, the East Iceland Arctic Current, turns to the east and forms a portion of the counterclockwise circulation in the southern part of the Norwegian Sea.

No velocities are entered on the figure, but within the Norwegian Current velocities up to 30 cm/sec, or about 0.6 knots, occur, and within the East Greenland Current, where the water moves fastest directly off the shelf, velocities of 25 to 35 cm/sec are encountered. These values are based partly upon computation from the distribution of density and partly upon direct current measurements.

In the North Sea a counterclockwise circulation is also present, as has been demonstrated not only by the distribution of temperature and salinity but also by the results of large-scale experiments with drift bottles. The details of the currents are much more complicated than is shown in the figure, and several smaller but permanent eddies appear to be present. In the straits between Sweden and Denmark the surface current is directed in general from the Baltic to the North Sea, but along the bottom the water flows into the Baltic. In the Baltic and the adjacent gulfs the currents are so much governed by local wind conditions that no generalization is possible.

In the North Polar Sea the surface currents are also greatly influenced by local winds. An independent current appears to be present only to the north of Spitsbergen and to the northeast of Greenland, where the surface waters flow south, feeding the East Greenland Arctic Current. The greater part of the Atlantic water that reaches Spitsbergen as the northern branch of the Norwegian Current submerges below the Arctic surface water and spreads as a warm intermediate layer over large parts of the Polar Sea.

In the Barents Sea a counterclockwise circulation prevails, with relatively warm water of Atlantic origin on the southern side and Arctic water on the northern, and with numerous eddies in the central portion. In the other marginal areas of the Polar Sea, the Kara Sea, the Laptev Sea, the North Siberian Sea, and the Chukotsk Sea, the currents are mainly determined by local winds and, in summer, by the discharge of large quantities of fresh water from the Siberian rivers or the Yukon River, but details of the currents are little known.

The Norwegian Current and its branches are subject to variations that are related to other phenomena. Helland-Hansen found that in 1929 the Atlantic water flowing into the Norwegian Sea was not only warmer and of higher salinity but that the volume was also greater. He points out that such an increased inflow must have had far-reaching consequences. Owing to the time used by the water to reach the Barents Sea, the effect of the large amount of warm water should have appeared in the Barents Sea two years after the water was observed off southwestern Norway, and the inflow of warmer water should have led to an increase of the ice-free areas in spring. In agreement with this reasoning the ice-free areas in the Barents Sea east of 20°E in May, 1929, comprised 330 km^2, and in May, 1931, they comprised 710 km^2. The causes of such fluctuations are not known, nor is it known whether these fluctuations are periodic in character, and the same applies to fluctuations which occur in other regions of well-defined currents.

THE LABRADOR SEA AND BAFFIN BAY. The surface currents of the Labrador Sea are shown in fig. 45. The outstanding features are the West Greenland Current, which flows north along the west coast of Greenland, and the Labrador Current, which flows south off the coast of Labrador. Part of the West Greenland Current turns around when approaching Davis Strait and joins the Labrador Current, whereas part continues into Baffin Bay, where it rapidly loses its character as a warm current. Along the west side of Baffin Bay the Arctic water flows south, having been partly reenforced by currents carrying Arctic water through the sounds between the islands to the west of Greenland. The central areas of the Labrador Sea and Baffin Bay both appear as areas in which numerous eddies occur, but nothing is known as to the permanency of the details shown in the picture. The similarity between the Labrador Sea and the Norwegian Sea is striking, but it should be emphasized that, whereas the Norwegian Sea is in communication with the Atlantic Ocean in the upper 600 m only, so that the deep water is shut off, the Labrador Sea is in communication with the Atlantic at all depths, and the deep water can flow freely to the south.

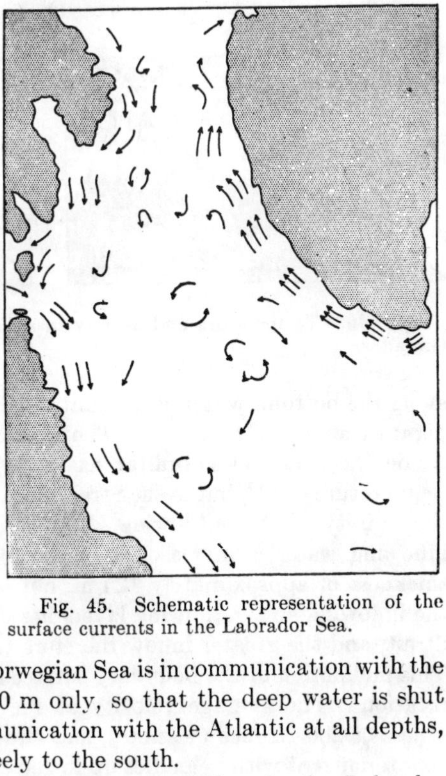

Fig. 45. Schematic representation of the surface currents in the Labrador Sea.

According to Smith, Soule, and Mosby the inflow in the Labrador Sea amounts to 7.5 million m^3/sec, and the outflow along the coast of Labrador amounts to 5.6 million m^3/sec, both values referring to flow above a depth of 1500 m. From these figures, Smith *et al* conclude that approximately 1.9 million m^3/sec sink and flow out from the Labrador Sea as deep water. This flow to the south has an important bearing on the entire deep-sea circulation of the Atlantic Ocean, as will be shown later.

THE MEDITERRANEAN AND THE BLACK SEA. The exchange of water between the North Atlantic Ocean and the Mediterranean takes place through the Strait of Gibraltar, which is noted for its strong currents. The bottom topography of the Strait is complicated, because in its shallow portion two or possibly three ridges are present with sill depths of about 320 m. The character of the water masses passing in and out through the Strait of Gibraltar is illustrated in fig. 46, which shows isohalines and isotherms in a longitudinal section through the Strait in the months of May, June, and July.

In the upper layers, Atlantic water of a salinity somewhat higher than 36.00°/₀₀ and of a temperature higher than 13° flows in, whereas,

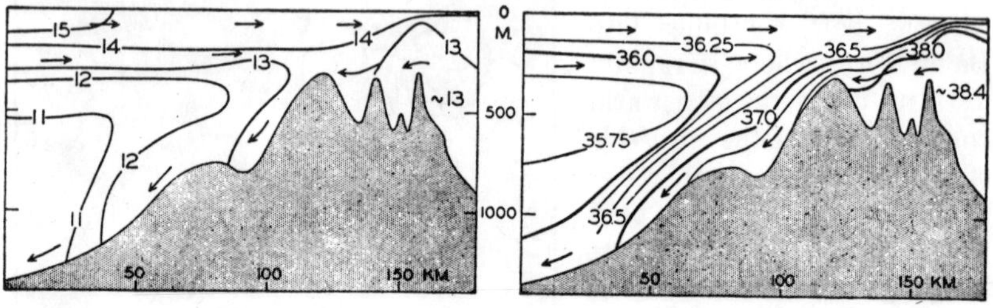

Fig. 46. Temperature and salinity in a vertical section through the Strait of Gibraltar.

along the bottom, water of a salinity higher than 37.00°/₀₀ and of a temperature about 13° flows out. The deeper water in the Mediterranean Sea inside the Strait of Gibraltar has a salinity of about 38.40°/₀₀ and a temperature of 13°, but in the Strait intensive mixing takes place by which the salinity of the outflowing water is greatly reduced and that of the inflowing water is increased. In the Strait the inflowing water has a thickness of approximately 125 m, but the boundary surface separating the inflowing and outflowing layers lies deeper on the African side of the Strait, and the greater inflow therefore takes place on the southern side. This inclination of the boundary surface is due to the effect of the earth's rotation. The average velocity at the surface in some localities is in excess of 200 cm/sec (4 knots), and close to the African coast a narrow countercurrent with velocities up to 100 cm/sec (2 knots) is often encountered. Superimposed on the currents carrying water in and out of the Mediterranean Sea are strong tidal currents that greatly reduce the inflow when they are directed from the Mediterranean Sea to the Atlantic Ocean, and greatly increase the inflow when they are directed from the Atlantic to the Mediterranean. The inclination of the boundary surface probably varies with the speed of the inflowing current, and great vertical oscillations of tidal period therefore take place.

The average velocity of the total inflow, according to measurements on Danish expeditions, is approximately 100 cm/sec (2 knots). By means of this value, Schott has computed the inflow to be approximately

1.75 million m³/sec. The average salinity of the inflowing water is about 36.25°/₀₀, and that of the outflowing water is about 37.75°/₀₀. With these values, one obtains an outflow of 1.68 million m³/sec, and the difference between inflow and outflow, 70,000 m³/sec, represents the excess of evaporation over precipitation and run-off. A fraction of this excess is made up, however, by a net inflow from the Black Sea, amounting to 6500 m³/sec (p. 180).

The exchange of water through the Strait of Gibraltar presents a good example of how water from one region can be transformed by external influences and return as a different type of water—in this case as water of high salinity. The rapidity of the exchange is illustrated by stating that the inflow and outflow are sufficient to provide for a complete renewal of the Mediterranean Water in about seventy-five years.

Using the figures that Schott gives for precipitation and runoff, one arrives at the water budget of the Mediterranean Sea proper, which is summarized in table 20. From this it appears that the total evaporation

TABLE 20

WATER BUDGET OF THE MEDITERRANEAN SEA

Gains	m³/sec	Losses	m³/sec
Inflow from the Atlantic Ocean	1,750,000	Outflow to the Atlantic Ocean	1,680,000
Inflow from the Black Sea	12,600	Outflow to the Black Sea	6 100
Precipitation	31,600	Evaporation	115,400
Run-off	7 300		
	1,801,500		1,801,500

amounts to 115,400 m³/sec, corresponding to an annual evaporation of 145 cm, which is in fair agreement with the annual evaporation in these latitudes according to observations and computations (fig. 13, p. 68). As one might expect, the evaporation from the Mediterranean Sea is somewhat greater than that from the open ocean in the same latitude, which is about 110 cm a year.

In fig. 47 a schematic picture is given of the surface currents and the flow of the Intermediate Water (according to Nielson). Both surface currents and intermediate currents have a tendency to circle the different areas in a counterclockwise direction.

The surface waters of the Black Sea flow out through the Strait of Bosporus and through the Dardanelles, and Mediterranean Water flows in along the bottom. Intensive mixing takes place in these narrow straits, and the salinity of the inflowing water is therefore reduced from more than 38.50°/₀₀ at the entrance of the Dardanelles to between 35.00 and 30.00°/₀₀ where the bottom current enters the Black Sea at the northern end of the Strait of Bosporus. Similarly, the salinity of the outflowing water is increased from about 16.00°/₀₀ where it enters the Strait of Bosporus to nearly 30.00°/₀₀ where it leaves the Dardanelles. The outflowing and inflowing water masses are separated by a well-defined layer of transition that oscillates up and down according to the contours of the bottom. The currents through the Strait of Bosporus can well be compared to two rivers flowing one above the other in a river bed that has a width of about 4 km and a depth of 40 to 90 m.

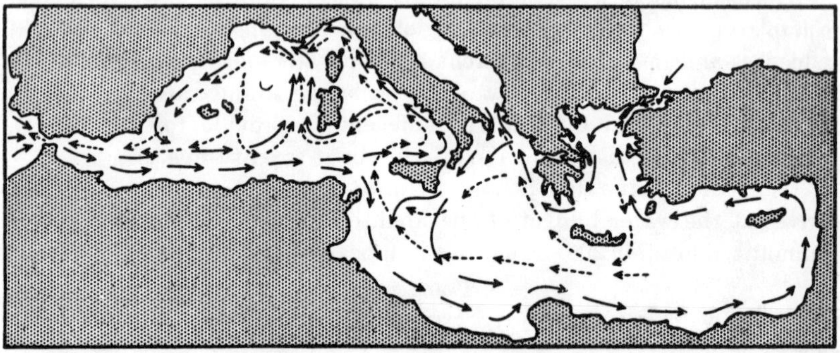

Fig. 47. Surface currents (solid arrows) and currents at intermediate depths (dashed arrows) in the Mediterranean Sea (after Nielsen).

Appreciable water masses pass through the Strait. According to current measurements and other observations by A. Merz, the most probable values of the outflow and inflow through the Strait of Bosporus are 12,600 m^3/sec and 6100 m^3/sec, respectively. (The Mississippi River carries, on an average, 120,000 m^3/sec.) The difference between inflow and outflow, 6500 m^3/sec, represents the excess of precipitation and run-off over the evaporation. Precipitation and run-off have been estimated at 7600 and 10,400 m^3/sec, respectively, and the evaporation should therefore amount to 11,500 m^3/sec, or 354 km^3/year. The area of the Black Sea is 420,000 km^2, and the above value therefore corresponds to an evaporation of 84 cm per year, in good agreement with observed and computed values for this latitude (p. 68).

The Caribbean Sea and the Gulf of Mexico. The surface currents in spring are shown in fig. 48. A strong current passes through the Caribbean Sea and continues with increased speed through the Yucatan Channel. There it bends sharply to the right and flows with great velocity out through the Straits of Florida. On the flanks of the main current, numerous eddies are present, of which the one in the wide bay between Nicaragua and Colombia and the one between Cuba and Jamaica are particularly conspicuous. In the Gulf of Mexico, several large eddies exist, all of which appear to be semipermanent features, and their locations seem to be determined by the contours of the coast and the configuration of the bottom.

The presentation of the currents in fig. 48 is based on ships' observations. When attempting a calculation of currents by means of numerous *Atlantis* data from the Caribbean Sea, Parr found that the flow is not directed parallel to the contours of the isobaric surfaces, but that between the Lesser Antilles and the Yucatan Channel the current flows uphill. Parr suggested that this feature may be due to a piling up of water in

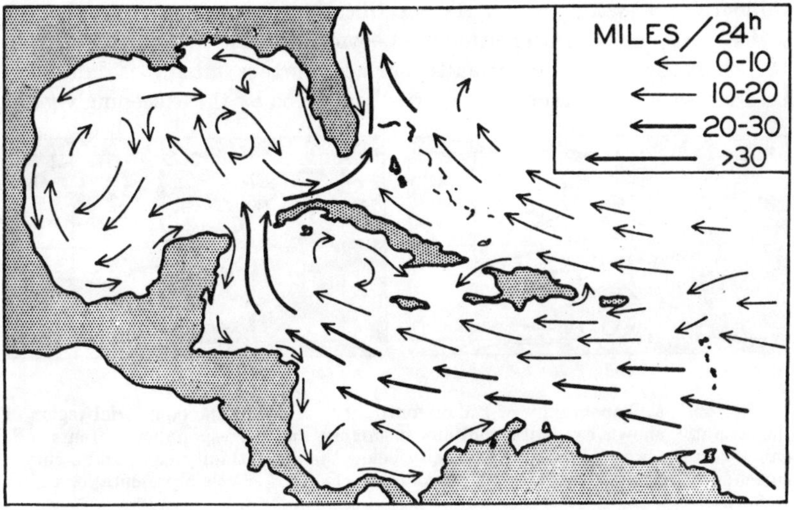

Fig. 48. Surface currents in spring in the Caribbean Sea and the Gulf of Mexico (after Dietrich).

front of the narrow Yucatan Channel that was caused by the stress exerted on the surface by the prevailing easterly winds (p. 122). This idea was further examined by Sverdrup, who concluded that the piling up of the surface water can be fully explained as the effect of winds blowing with an average velocity of about 10 m/sec, which agrees well with the observed values in spring. A further consequence of this piling up is that in the Gulf of Mexico a sea level is maintained higher than that along the adjacent coast of the United States facing the Atlantic Ocean. At Cedar Keys, off the southwest tip of Florida, the average sea level is 19 cm higher than the average sea level at St. Augustine, Florida, on the east coast. This indicates that the prevailing winds over the Caribbean Sea produce a hydrostatic head which, according to Montgomery, may account for the major part of the energy of the Florida Current (p. 164).

Currents of the Equatorial Part of the Atlantic Ocean

In the equatorial region the water masses below a depth of 50 to 150 m are separated from the surface waters by a layer within which the density increases so rapidly with depth that it has the character of a discontinuity layer. The discontinuity layer is explained mainly as a result of heating due to the absorption of radiation from the sun and forced mixing due to wind within the water. Since this water comes from higher latitudes, it has a relatively low temperature. As the mixing penetrates to greater and greater depths, the difference in density between the warm surface waters and the colder subsurface strata is concentrated in a thinner and thinner layer within which the stratification becomes more and more stable. When the stratification has become so stable that the layer takes the character of a discontinuity surface, further mixing is inhibited, because the eddy conductivity is greatly reduced by the exceedingly great

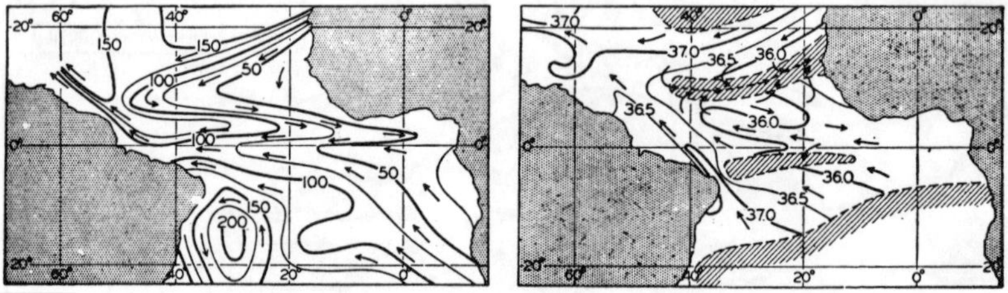

A B

Fig. 49. A. Topography of the discontinuity surface in the equatorial region of the Atlantic, shown by depth contours in meters, and corresponding currents. B. Salinity in the layer of maximum salinity below the discontinuity layer and assumed currents. Regions without salinity maximum shaded. (Both representations after Defant.)

stability (p. 20), and the amounts of heat that are conducted downward become so small that they are carried away by weak currents. Additional absorption of solar energy is used mainly for evaporation; the final temperature that is attained will depend upon the rate of evaporation, and is therefore determined by the interaction between the atmosphere and the ocean.

In a given locality the thickness of the upper warm layer depends not only upon the intensity of mixing but also upon the character of the circulation in the top layers, because in the presence of currents the discontinuity surface cannot remain horizontal, but must slope (p. 103). On the basis of the *Meteor* data and all other observations available from the tropical parts of the Atlantic Ocean, Defant has been able to construct a chart showing the topography of the discontinuity surface between latitudes 25°N and 25°S (fig. 49). In the presence of a sloping discontinuity surface the flow of the water above the surface, relative to the underlying water masses, must, in the Northern Hemisphere, be in such direction that the surface sinks to the right of an observer looking in the direction of flow, and in the Southern Hemisphere such that the surface sinks to the left. On the basis of these rules, arrows have been entered showing the direction of flow, which, in general, agrees well with the observed average surface currents in the tropical regions of the Atlantic.

This picture shows a gyral in the South Atlantic Ocean and demonstrates the complicated character of the currents near the Equator, where a countercurrent toward the east, the Equatorial Countercurrent, is imbedded between the equatorial currents of the two hemispheres.

Evidentiy the countercurrent is related to the distribution of mass, as is demonstrated by the slope of the discontinuity surface. In a profile from south to north the discontinuity surface rises toward the Equator, reaches a maximum elevation at the Equator, and drops toward a minimum in 2° to 3°N latitude at the northern boundary of the South Equatorial Current. Between 2° to 3°N and 6° to 8° N the discontinuity surface rises toward the north, reaching a second and more pronounced maximum in the latter latitude. This is the region of the countercurrent, and to the north of the second maximum, where the surface slopes downward again, the North Equatorial Current is found. This distribution of mass was first recognized by means of the *Carnegie* observations in the Pacific (fig. 52, p. 194).

The dynamics of the countercurrent has recently been discussed by Montgomery, who writes:

> The trade winds, by continually exerting a westward stress on the sea surface, produce a westward ascent of sea level along the Equator. This slope amounts to about 4 cm per thousand km, and the accompanying pressure gradient extends down to about 150 m in the Atlantic Ocean. The Equatorial Countercurrents are found in the doldrums and apparently result simply as a downslope flow in this zone where the winds maintaining the slope are absent.

The ascent of sea level from east to west is due to a greater accumulation of light surface water along the coasts of South America. Fig. 49 shows that such accumulation exists, because the thickness of the homogeneous surface layer above the discontinuity surface increases from less than 40 m in the east to about 140 m in the west. Above a depth of 150 m the isosteric surfaces, therefore, slope downward from east to west, but below 150 m they are practically horizontal. Accordingly, Montgomery and Palmén find that at the Equator the dynamic height of the sea surface, referred to the 1000-decibar surface, is 14 dyn/cm greater in the west than in the east, but that the 150-decibar surface is parallel to the 1000-decibar surface. The countercurrent is therefore a very shallow current that is confined to the surface layer above the discontinuity.

A swift current that is embedded between water masses moving in the opposite direction must be subject to a considerable retardation because of friction. A certain amount of energy is therefore needed for maintaining such a current, and this energy, according to Montgomery and Palmén, is derived from the trade winds, which maintain the slope of the sea surface. Montgomery and Palmén assume that the retarding friction may be due to lateral mixing with the westward-flowing equatorial currents, and estimate the coefficient of horizontal eddy viscosity to be 7×10^7 g/cm/sec. This concept leads to the important conclusion that on both sides of the countercurrent the equatorial currents are subjected to great lateral frictional stresses that are directed opposite to the horizontal stresses exerted by the trade winds on the surface. Similar lateral stresses must be directed opposite to the general flow along the continental boundaries of the two large counterclockwise gyrals of the two hemispheres, and the torque exerted by these stresses perhaps balances the torque exerted by the stress of the wind on the sea surface. The

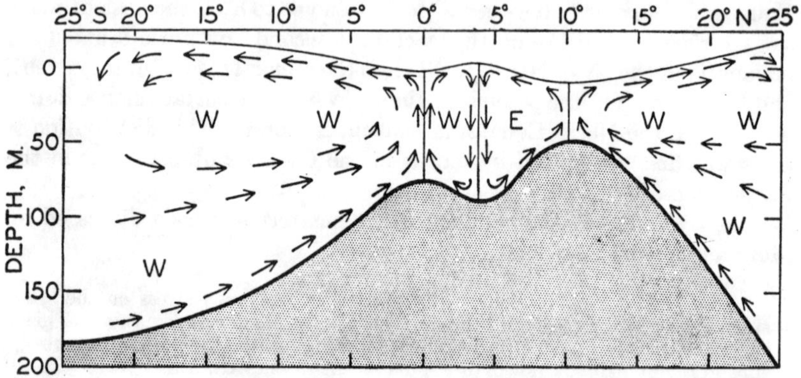

Fig. 50. Schematic representation of the vertical circulation within the equatorial region of the Atlantic. The main direction of the currents is indicated by the letters *W* and *E*. The water below the discontinuity surface, which is supposed to be at rest, is shaded.

lateral stresses between the equatorial currents and the countercurrent probably contribute very materially to the total torque, and the countercurrent represents, therefore, a dynamically important link in the entire system of ocean currents.

Within the countercurrent and the adjacent equatorial currents the frictional forces must lead to a transport of water across the isolines—that is, to a transverse circulation. This question was examined theoretically by Defant, who considered friction due to vertical mixing, but the conclusions are probably also applicable if the friction is due to lateral mixing. Defant's results are shown schematically in fig. 50, according to which four "cells" are present, representing gyrals with horizontal axes, neighboring gyrals rotating in opposite directions. Within the southern cell the water sinks in the region of the Tropical Convergence and rises at the Equator, and within the next cell, located between the Equator and the southern boundary of the countercurrent, the water rises at the Equator and sinks at the boundary of the countercurrent. Within the countercurrent the water rises at the northern boundary and sinks at the southern, and within the northern cell sinking motion takes place at the

Tropical Convergence and rising motion takes place at the northern boundary of the countercurrent. The data from the Atlantic Ocean indicate the existence of these cells, and the Carnegie section across the Pacific Countercurrent (fig. 52, p. 194) demonstrates their presence convincingly. As a consequence of these transverse circulations the northern boundary of the countercurrent and of the Equator represent lines of *divergence*, whereas the southern boundary of the countercurrent is a line of *convergence*, and individual water masses carried by the different currents follow complicated spiral-like trajectories.

In summer, when the countercurrent is best developed, the effect of the divergence at the Equator appears on the charts of surface temperatures as a tongue of low temperature, but no effect of the divergence at the northern boundary of the countercurrent is visible (chart 1).

Currents of the South Atlantic Ocean

The most outstanding current of the South Atlantic Ocean is the Benguela Current, which flows north along the west coast of South Africa and is particularly conspicuous between the south point of Africa and latitude 17° to 18°S. In agreement with the dynamics of currents in the Southern Hemisphere the denser water of low temperature is found on the right-hand side of the current—that is, close to the African coast. Under the influence of the prevailing southerly and southeasterly winds the surface layers are carried away from the coast, and upwelling of water from moderate depths takes place in most seasons of the year. As a consequence of this upwelling a band of water of low temperature and of relatively low salinity is found along the coast extending out to a distance of approximately 200 km.

Proceeding toward the Equator, the Benguela Current gradually leaves the coast and continues as the northern portion of the South Equatorial Current, which flows toward the west across the Atlantic Ocean between latitudes 0° and 20°S. In the charts which show the distribution of surface temperatures, the tongue of low temperature along the Equator is due not only to the cold waters of the Benguela Current, but also to the divergence along the Equator. The South Equatorial Current partly crosses the Equator, continuing into the North Atlantic Ocean, and it partly turns to the left and flows south along the South American coast, where it shows up as a tongue of water of high temperature and high salinity, the Brazil Current. Close to the coast of Argentina a branch of the Falkland Current extends from the south to about latitude 30°S, carrying water of lower temperature and lower salinity. Thus the most conspicuous feature of the currents of the South Atlantic is represented by the counterclockwise gyral with the cold Benguela Current on the eastern side, the warm Brazil Current on the western side, the South Equatorial Current flowing west on the northern side, and the South Atlantic Current flowing east on the southern side. It is a system of shallow currents, because the entire circulation takes place above the Antarctic Intermediate Water, and near the Equator it is probably limited to a depth of less than 200 m.

From the data of the *Meteor* expedition, it is possible to arrive at some quantitative conclusions. Calculations of the transport through vertical sections between South Africa and South America (p. 116) lead to the result that the transport above a depth of 4000 m is directed toward the north, but this is obviously impossible, because the *net* transport through such a section must be nearly zero. In calculating the transport, it is therefore necessary to assume a layer of no motion at some intermediate depth above which the transport will be directed toward the north and below which it will be directed toward the south. The *Meteor* profiles to the south of latitude 20°S lead consistently to the result that the intermediate layer of no motion is found at a depth of about 1300 m—that is, at the boundary between the Antarctic Intermediate Water and the Deep Water. Thus, the intermediate water and the upper water move, on the whole, to the north, whereas the deep water moves south, but close to the bottom there is a flow of Antarctic Bottom Water toward the north. The net transport to the south below 1300 m amounts approximately to 15 million m^3/sec.

TABLE 21

TRANSPORT OF WATER ACROSS LATITUDE 30°S AND ACROSS THE EQUATOR

Latitude	Water mass	Current	Transport in million m^3/sec toward	
			North	South
30°S	Upper water	Benguela Current	16	
		Brazil Current		10
		Central part of South Atlantic gyral	7	7
	Intermediate water		9	
	Deep water			18
	Bottom water		3	
0°	Upper water		6	
	Intermediate water		2	
	Deep water			9
	Bottom water		1	

The transport through a vertical section in the South Atlantic can be studied in greater detail, and approximate values can be computed of the amounts of the different types of water that are transported to the north and to the south, taking the continuity of the system into account. In table 21 are shown results of such computations as referred to a section in 30°S. The table also contains values for the transport across the Equator, but these have been obtained in a different manner. The value of a northward transport of about 6 million m^3/sec of upper water is mainly based on a comparison between the waters of the Caribbean Sea and the Sargasso Sea, and the values of the northward transports of intermediate and bottom water are estimated.

The approximate correctness of the figures can be checked by considering that the net transport of salt must be zero. A calculation of the net transport of salt requires knowledge of the velocity distribution within the different parts of the current system, but a rough computation based on average salinity values confirms the above conclusions. We shall have to return to some of these matters when discussing the deep-water circulation, but here it will be emphasized that a large quantity of South Atlantic Central and Intermediate Water crosses the Equator and enters the North Atlantic Ocean, where it exercises an influence on the character of the waters along the coast of South America, in the Caribbean Sea, and in the Gulf of Mexico.

Currents of the Indian Ocean

In the southern part of the Indian Ocean a great anticyclonic system of currents comparable to the corresponding systems of the North and South Atlantic Ocean appears to prevail, except that it is subjected to greater annual variations. Between South Africa and Australia the current is directed in general from west to east. In the southern summer the current bends north before reaching the Australian Continent and is joined by a current that flows from the Pacific to the Indian Ocean to the south of Australia. In winter the current appears to reach to Australia and in part to continue toward the Pacific along the Australian south coast. To the north of 20°S the South Equatorial Current flows from east to west, reaching its greatest velocity during the southern winter, when the southwest monsoon over the northern part of the ocean represents a direct continuation of the southeast trade winds on the southern side of the Equator. In this season the current is reinforced by water from the Pacific Ocean that enters the Indian Ocean to the north of Australia, but in the southern summer the flow to the north of Australia is reversed.

In both seasons of the year, part of the South Equatorial Current turns south along the east coast of Africa and feeds the strong Agulhas Stream. To the south of lat. 30°S the Agulhas Stream is a well-defined and narrow current that extends less than 100 km from the coast. Corresponding to a flow toward the south in the Southern Hemisphere, the coldest water is found inshore, and the sea surface rises when departing from the coast. Off Port Elizabeth the rise amounts to about 29 cm in a distance of 100 km. To the south of South Africa the greatest volume of the waters of the Agulhas Stream bends sharply to the south and then toward the east, thus returning to the Indian Ocean by joining the flow from South Africa toward Australia across the southern part of that ocean, but a small portion of the Agulhas Stream water appears to continue into the Atlantic Ocean. Owing to the reversal of the direction of the main current to the south of Africa, numerous eddies develop tha' result in a highly complicated system of surface currents which probably is subjected to considerable variations during the year and to variations from one year to another.

To the north of lat. 10°S the surface currents of the Indian Ocean, which are probably nearly identical with the currents above the tropical discontinuity surface, vary greatly from winter to summer owing to the different character of the prevailing winds. During February and March, when the northwest monsoon prevails, the North Equatorial Current is well developed, and an Equatorial Countercurrent with its axis in approximately 7°S is present. Along the African east coast, between the Gulf of Aden and lat. 5°S, the current is directed toward the south. In August and September, when the southwest monsoon blows, the North Equatorial Current disappears and is replaced by the Monsoon Current, which flows from west to east. Along the coast of east Africa, where the current is directed north from lat. 10°S, water of the Equatorial Current crosses the Equator, and considerable upwelling takes place off the Somali coast. The Equatorial Countercurrent does not appear to be present in this season.

Nothing is known as to the motion of the water below the tropical discontinuity in the northern part of the Indian Ocean. The character of the water indicates that no strong currents exist and that only a sluggish flow takes place.

In the southern part of the Indian Ocean the Antarctic Intermediate Water probably flows north, but the flow must be less well-defined than the corresponding flow in the South Atlantic Ocean, because in the Indian Ocean the Antarctic Intermediate Water loses its typical characteristics in a shorter distance from the Antarctic Convergence. The probable flow of the deep water will be discussed later.

The data from the Indian Ocean are too scanty to permit many quantitative calculations as to the amount of water carried by the different branches of the current. The only reliable figure that is available is found in Dietrich's study of the Agulhas Stream, which transports a little more than 20 million m^3/sec. A transport map similar to the one for the North Atlantic Ocean cannot be constructed.

THE RED SEA. The exchange of water between the Red Sea and the adjacent parts of the ocean takes place through the Suez Canal and through the Strait of Bab-el-Mandeb. According to Vercelli and Thompson, this exchange is subject to a distinct annual variation that is related to the change in the direction of the prevailing winds in winter and summer. In winter, when south-southeast winds blow in through the Strait of Bab-el-Mandeb, the surface layers are carried from the Gulf of Aden into the Red Sea, and at greater depths highly saline Red Sea Water flows out across the sill. In summer, when north-northwest winds prevail, the surface flow is directed out of the Red Sea, and at some intermediate depth water having a lower salinity and a lower temperature than the outflowing surface water flows in from the Gulf of Aden. At still greater depths, highly saline Red Sea Water appears to flow out over the sill, but it is probable that this outflow is much less than the outflow in winter. On the basis of direct measurements of currents at anchor stations, Vercelli found that in winter the average inflow amounts to approximately 0.58 million m^3/sec, whereas the outflow of Red Sea water amounts to approximately 0.48 million m^3/sec. No measurements are available for summer, but it is estimated that the average annual outflow is between 0.3 and 0.4 million m^3/sec—that is, approximately one sixth of the amount that flows out through the Strait of Gibraltar. This conclusion is in agreement with the fact that the Red Sea Water is of less importance in the Indian Ocean than the Mediterranean Water is in the Atlantic Ocean.

Currents of the South Pacific Ocean

The only major current of the South Pacific Ocean that has been examined to some extent is the Peru Current. In accordance with the nomenclature proposed by Gunther, the name "Peru Current" will be applied to the entire current between the South American Continent and the region of transition toward the eastern South Pacific Central region. The part of the current which is close to the coast will be called the Peru Coastal Current, whereas the part which is found at greater distances will be called the Peru Oceanic Current.

The origin of the Peru Current has to be sought in the subantarctic region, where part of the Subantarctic Water that flows toward the east across the Pacific Ocean is deflected toward the north when it approaches the American Continent. The total volume of water of the current does not appear to be very great. On the basis of a few *Discovery* stations, it is found that the transport lies somewhere between 10 and 15 million m^3/sec, and this figure includes transport of the upper water layers and of Antarctic Intermediate Water. The western limit of the current appears to be diffuse and cannot be well established on the basis of the available data, but it is probable that the current extends to about 900 km from the coast in 35°S. The northern limits, according to Schott, can be placed a little south of the Equator, where the flow turns toward the west.

Owing to the width of the current and the small transport, its velocities are quite small. This must be true particularly in the case of the Peru Oceanic Current, which, however, is little known. Within the better known Peru Coastal Current the upwelling represents a very conspicuous feature. The upwelling is caused by the southerly and south-southeast winds that prevail along the coasts of Chile and Peru and carry the warm and light surface waters away from the coast, resulting in cold water being drawn from moderate depths (40 to 300 m) toward the surface.

According to Schott and to Gunther, the most active upwelling occurs in certain regions separated by regions in which the upwelling is less intense. Both authors recognize four such regions between lat. 3°S and 33°S, but they do not agree on the extent of the different regions, probably because the regions are not absolutely fixed or because the locations ascribed to the different regions may depend upon the available data. Gunther has particularly examined the two northern regions, where the most intense upwelling occurs in 5°S and 15°S, respectively, and has shown that the surface temperatures in the winter of 1931 (June to August) indicate the existence of two tongues of warm water that approach the coast to the south of the regions of intense upwelling. The upwelled water, on the other hand, leaves the coast as tongues of cold water, and thus the distribution of surface temperatures shows alternate tongues of warm and cold water. Schott's analysis and other observations indicate that the locations of these tongues do not vary much from one year to another, and the tongues must therefore be either permanent or recurrent. Gunther interprets these tongues as demonstrating the existence of swirls off the coast, assuming that within one branch of the swirl upwelled water moves out and that within another branch oceanic water moves toward the coast.

In early winter, April to June, the shoreward-directed branch of the northern swirl is well developed and carries water of high temperature in toward the coast in latitudes 9° to 12°S. It may even appear as an inshore warm current that brings great destruction to the animal life of the coastal waters. In the discussion of the California Current, it will be shown that this current is similarly characterized by a series of swirls on the coastal side of the current, and that in November to February, when there is practically no upwelling, a warm countercurrent flows to the north along the coast.

At the northern boundary of the Peru Coastal Current, certain characteristic seasonal changes take place. During the northern summer the Peru Coastal Current extends just beyond the Equator, where it converges with the Equatorial Countercurrent, the waters of which in summer turn mainly toward the north. In winter the countercurrent is displaced further to the south, and part of the warm but low-salinity water of the countercurrent turns south along the coast of Ecuador, crossing the

Equator before converging with the Peru Coastal Current. The warm, south-flowing current along the coast is known as "El Niño," and is a regular phenomenon in February and March, but the southern limit lies mostly only a few degrees to the south of the Equator. Occasionally, major disturbances occur that appear to be related to changes in the atmospheric circulation. In disturbed years, such as in 1891 and in 1925, the El Niño extends far south along the coast of Peru, occasionally reaching past Callao, in 12°S. According to Schott, the duration of the El Niño period of 1925 was as follows:

Off Lobitos	lat. 4°20′S	Jan. 20–April 6	76 days
Off Puerto Chicana	" 7°40′S	Jan. 30–April 2	63 days
Off Callao	" 12°20′S	March 12–27	15 days
Off Pisco	" 13°40′S	March 16–24	8 days

These figures show that the warm surface waters of the equatorial area slowly penetrated to the south, but withdrew much more rapidly, because the time interval between the appearance of the warm water off Lobitos and off Pisco was 44 days, whereas the time interval between the disappearance of the warm water at the two localities was only 13 days. The surface temperature of the water in March, 1925, was up to 7° above the average, as shown in the following compilation:

Locality	Average temperature in March (°C)	Temperature in March, 1925 (°C)
Lobitos	22.2	27.3
Puerto Chicana	20.3	26.9
Callao	19.5	24.8
Pisco	19.0	22.1

Details as to the surface salinity are not available, but normally the surface salinity between 5°S and 15°S is above 35.00°/₀₀, whereas the waters of the El Niño have a salinity between 33.00 and 34.00°/₀₀.

The extreme development of the El Niño leads to disastrous catastrophes of both oceanographic and meteorological character. The decrease of the El Niño temperature toward the south indicates that the waters are mixed with the ordinary cold coastal waters, and during this process the organisms in the coastal current, from plankton to fish, are destroyed on a wholesale scale. Dead fish later cover the beaches, where they decompose and befoul both the air and the coastal waters. So much hydrogen sulphide may be released that the paint of ships is blackened, a phenomenon known as the "Callao painter." More serious is the loss of food to the guano birds, many of which die of disease or starvation or leave their nests, so that the young perish, bringing enormous losses to the guano industry. The meteorological phenomena that accompany the El Niño are no less severe. Concurrent with a shift in the currents a shift of the tropical rain belt to the south takes place. Thus, in March, 1925, the precipitation at Trujillo, in 8°S, amounted to 395 mm, as compared to an average precipitation in March of the 8 preceding years of only 4.4 mm. These terrific downpours naturally cause damaging floods and erosion. In the 140 years from 1791 to 1931, 12 years were characterized by excessive rainfall at Piura, in lat. 5°S, and 21 years by moderate rainfall which were, however, greatly in excess of the average. During the remaining nearly 100 years the rainfall was close to nil. A greater development of El Niño is therefore not an uncommon phenomenon, but on an average the catastrophic developments appear to occur once in 12 years. The records reveal no periodicity, because the interval between two disastrous years varies from 1 year to 34 years.

The El Niño is not the only current that brings warm water to the coast of Peru with subsequent destruction of the organisms near the coast. High temperatures off the coast appear to be an annual occurrence in the months of April to June in about lat. 9° to 12°S—that is, at and to the north of Callao. These high temperatures are due, as was pointed out by Gunther, to the greater development of the warm branch of the northern swirl, and the water that approaches the coast is in this case offshore oceanic surface water of high temperature and relatively high salinity (p. 190). The disastrous effect on the marine organisms is very much milder than that caused by the El Niño, but is otherwise similar in character. It may lead to the killing of plankton and fish and to the migration of guano birds, but is ordinarily observed mainly by a change in the color of the coastal water and by development of hydrogen sulphide. Locally, these changes are known by the name of "aguaje," but this name is also used synonymously with "Callao painter." The approach of the oceanic water toward the coast is not accompanied by any disastrous meteorological conditions.

On leaving the coast, the waters of the Peru Current join the waters of the South Equatorial Current, which flows all the way across the Pacific toward the west, but is known only as far as surface conditions are concerned. The subsurface data are inadequate for computation of velocities and transports, and it is therefore not possible to give any numerical values. Some features of this current will, however, be dealt with when the currents of the equatorial regions of the Pacific are discussed, and it will be shown that the cold water along the Equator does not represent a continuation of the Peru Coastal Current, but is due to a divergence along the Equator within the South Equatorial Current.

The other currents of the South Pacific Ocean are even less known, but from the character of the water masses it appears that two current systems exist whose nature may be revealed by future exploration. One big gyral appears to be present in the eastern South Pacific, but in the western South Pacific annual variations are so great that in many regions the direction of flow becomes reversed, as is the case off the east coast of Australia. No chart of the transport by the current can be prepared.

Currents of the Equatorial Pacific

In the Equatorial Pacific a chart of the depth to the thermocline immediately gives an idea of the currents in the upper layers in the tropical region (fig. 51). If the motion of the water below the discontinuity

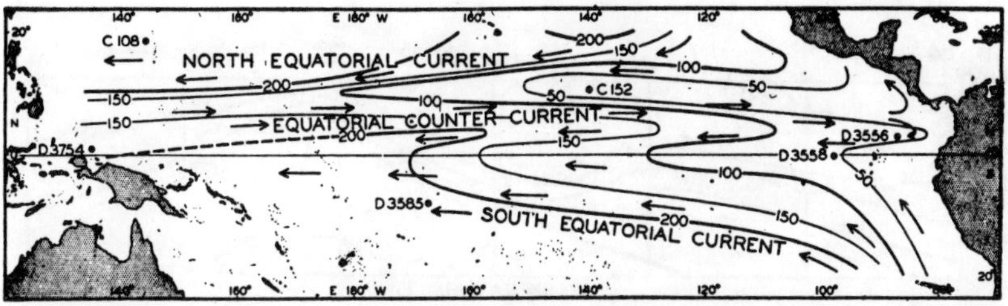

Fig. 51. Topography of the discontinuity surface in the equatorial region of the Pacific shown by depth contours in meters, and corresponding currents.

layer is small, the current in the upper layers must be related to the slope of the discontinuity layer in such a manner that in the Northern Hemisphere the discontinuity layer sinks to the right of an observer looking in the direction of the current and, in the Southern Hemisphere, sinks to the left (p. 103). In the figure, arrows have been entered on the basis of this rule showing the North and South Equatorial Currents flowing toward the west and between them the Equatorial Countercurrent flowing toward the east. The South Equatorial Current is present on both sides of the Equator and extends to about 5°N, but the North Equatorial Current remains in the Northern Hemisphere. Off South America the flow is directed more or less parallel to the coast line, turning gradually west when approaching the Equator.

The Equatorial Countercurrent is remarkably well developed in the Pacific Ocean, where, according to charts published by Puls in 1895, it is present at all seasons of the year, lying always in the Northern Hemisphere, but further away from the Equator in the northern summer. In this season the velocities of the current also appear to be higher, reaching values up to 100 cm/sec at the surface. The structure of the water masses was first demonstrated by the *Carnegie* section in approximately long. 140°W, which was obtained in October, 1929. Fig. 52 (top and center) shows the temperature and salinity in this section between the surface and 300 m. The figure brings out the great variation in the depth to the thermocline in a north-south direction, the presence of the surface layer of uniform temperature, and the tongue of maximum salinity, which, to the south of the Equator, lies somewhat above the thermocline. The bottom section shows the velocity distribution, which has been computed on the assumption of no motion at the 700-decibar surface. These computations are uncertain near the Equator, but the resulting picture is remarkably consistent. In this case the countercurrent lies between 3°N and 10°N, as indicated by the letters stating the direction in which the currents flow.

The maximum velocity at the surface, as computed from the *Carnegie* section (fig. 52), is a little over 50 cm/sec, or about 1 knot, in good

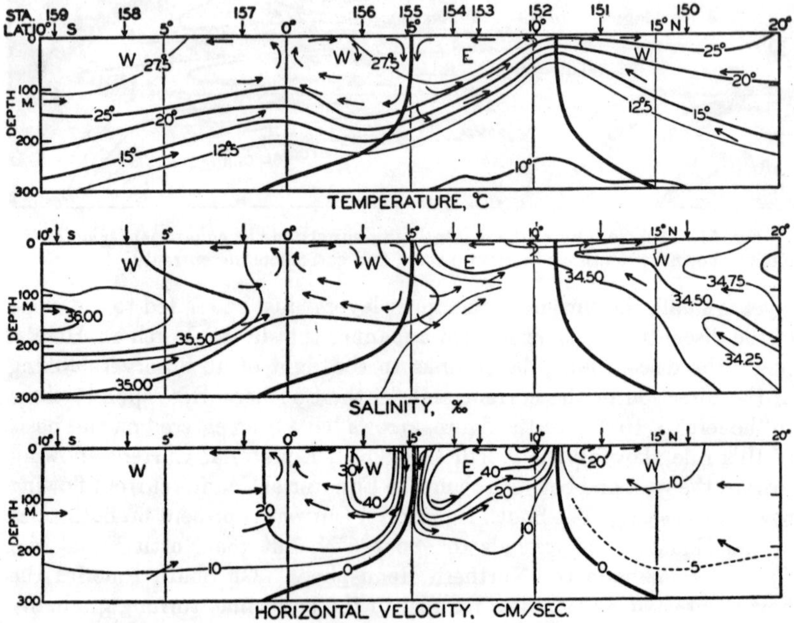

Fig. 52. Temperature, salinity, and computed velocity in a vertical section between 10°S and 20°N in the Pacific Ocean, according to the *Carnegie* observations. Arrows indicate direction of north-south flow. *E* indicates flow to the east, *W* shows flow to the west.

agreement with values reported from ships. According to the *Carnegie* section the countercurrent transports approximately 25 million m^3/sec to the east, and the volume transport of the countercurrent of the Pacific is therefore comparable to that of the Florida Current. The surface observations seem to indicate that the transport is somewhat less in the western part of the ocean and that water is drawn into the current as it crosses the Pacific Ocean.

Montgomery and Palmén have shown that the countercurrent in the Pacific Ocean, as well as that in the Atlantic Ocean, is maintained by the piling up of the light surface water against the western boundary of the ocean, leading at the Equator to a slope of the sea surface of 4.5×10^{-8}.

Within the countercurrent, descending motion takes place at the southern boundary and ascending motion at the northern boundary, and between the Equator and the countercurrent descending motion takes place at the boundary of the countercurrent and ascending motion at the Equator. Thus, two cells appear with divergence at the northern limit of the countercurrent and at the Equator, and with a convergence at the southern boundary of the countercurrent. This system, which appears so clearly in the *Carnegie* section, is quite similar to the one that Defant, on the basis of theoretical considerations, has derived for the countercurrent in the Atlantic Ocean (fig. 50, p. 184).

Owing to the convergence and the divergences the water does not flow due east or due west, but superimposed upon the major current there is a spiral motion which within the countercurrent carries water from the northern to the southern boundary at the surface and carries water in the opposite direction at depths between 50 and 200 m. Similarly, within the Equatorial Current between the Equator and the countercurrent the surface water moves from the Equator toward the countercurrent, but at depths between 100 and 150 m the water moves in the opposite direction. To the north of the countercurrent and to the south of the Equator, subsurface water moves toward the divergences at the surface, and this water must originate from the regions of the Tropical Convergences, which lie outside the section under consideration. According to Defant's estimate the maximum north-south component of velocity is not more than one fifth of the east-west component.

From the above discussion it is evident that the equatorial divergence is not related to the proximity of land and that the conditions met with in the Galapagos area and to the west do not simply represent a continuation of the conditions off the coast of Peru, as had been previously assumed.

The character of the countercurrent is complicated both at the origin of the current between New Guinea and the Philippines and at the termination against the American coast. Schott has shown that large season changes take place to the north of New Guinea, where from June to August the South Equatorial Current follows the north coast of New Guinea and converges sharply with the North Equatorial Current in about lat. 5°N, where the countercurrent begins, whereas from December to February part of the North Equatorial Current bends completely around off the southern islands of the Philippines, sending one branch toward the southeast along the north coast of New Guinea, and another branch, the countercurrent, toward the east.

Similarly, great seasonal variations occur in the Central American region, as is evident from the data presented on the U. S. Hydrographic Pilot Charts, but in this region the picture is complicated by numerous eddies whose locations appear to vary from one year to another. The only persistent features are that the countercurrent is in most months well developed between lat. 5° and 6°N and that the greater volume of water transported by the countercurrent is deflected to the north and northwest, where a strong current prevails off the coast of Central America. Another but weaker and much more irregular branch turns to the south. In the Gulf of Panama, large seasonal changes occur that are associated with the change in the prevailing wind direction.

Currents of the North Pacific Ocean

THE NORTH EQUATORIAL CURRENT. There is considerable similarity between certain currents of the North Pacific and of the North Atlantic Oceans, but there are also striking differences that are due mainly to the occurrence of large quantities of Subarctic Water in the North Pacific, as contrasted to the small amounts of that type in the North Atlantic. The Subarctic Water of the Pacific and the currents that carry Subarctic Water are present in the northern and eastern areas, and similarity to the North Atlantic is therefore found in the southern and western part of the ocean, where the North Equatorial Currents correspond to each other and where the Kuroshio corresponds to the Florida Current and the Gulf Stream.

The North Equatorial Current of the Pacific Ocean runs from east to west, increasing in volume transport because new water masses join the current from the north. The very beginning of the North Equatorial Current is found where the waters of the countercurrent turn to the north off Central America. To these water masses are later added the waters of the California Current, which have attained a relatively higher temperature and salinity, owing to heating and evaporation, and which have been mixed with waters of the tropical region. Between the American coast and the Hawaiian Islands, Eastern North Pacific Water is added to the equatorial flow, and to the west of the Hawaiian Islands a considerable addition of Western North Pacific Water appears to take place. No details are known as to the character of the current, and only a few isolated computations of velocities can be made. These indicate that one has to deal with a broad and relatively deep current within which the maximum velocities are mostly less than 20 cm/sec. A vertical transverse circulation is present with ascending motion at the northern boundary of the countercurrent and with descending motion at the Tropical Convergence. The waters therefore move in a spiral-like fashion, and the motion is probably highly complicated owing to the great distance between the regions of major divergence and convergence and to the presence of local divergences and convergences.

It is probable that before reaching the western boundary of the ocean the Equatorial Current begins to branch off mainly to the north, but also partly to the south, feeding the countercurrent. To the east of the

Philippine Islands a definite division of the current takes place, one branch turning south along the coast of the island of Mindanao and the larger branch turning north, following closely the east side of the northern Philippine Islands and the island of Formosa. The intensity of the flow to the south varies with the season, and the same probably is true regarding the flow to the north.

THE KUROSHIO SYSTEM. After passing the island of Formosa the warm waters continue to the northeast between the shallow areas of the China Sea and the submarine ridge on which the Riukiu Islands lie. On reaching lat. 30°N the current bends to the east and then to the northeast, following closely the coast of Japan as far as lat. 35°N. The name "Kuroshio" is particularly applied to the current between Formosa and lat. 35°N, but, in agreement with the nomenclature used when dealing with the currents of the Atlantic, and following Wüst, we shall apply the name "Kuroshio system" to all branches of the current system and shall introduce the following subdivisions:

1. The Kuroshio: The current running northeast from Formosa to Riukiu and then close to the coast of Japan as far as lat. 35°N.
2. The Kuroshio Extension: The warm current that represents the direct continuation of the current and flows nearly due east, probably in two branches, and can be traced distinctly to about longitude 160°E.
3. The North Pacific Current: The further continuation of the Kuroshio Extension that flows toward the east, sending branches to the south and reaching probably as far as long. 150°W.

To these three major divisions of the Kuroshio system can also be added the *Tsushima Current*, the warm current which branches off on the left-hand side of the Kuroshio and enters the Japan Sea following the western coast of Japan to the north, and the *Kuroshio Countercurrent*, which represents part of a large eddy on the right-hand side of the Kuroshio.

THE KUROSHIO. The great similarity between the Kuroshio and the Florida Current has been pointed out by Wüst. Wüst compares the Kuroshio between the Riukiu Islands and the continental shelf to the current through the Caribbean Sea, and the flow of the Kuroshio across the Riukiu Ridge in lat. 30°N to the flow through the Straits of Florida. This part of the Kuroshio, as well as the current to the south of Shionomisaki in lat. 33°N, long. 135°W, has been studied by Japanese oceanographers who have conducted current measurements and have occupied a large number of oceanographic stations. According to Koenuma, the current between Formosa and the southern Riukiu Islands reaches to a depth of about 700 m, and the maximum velocity near the surface is 89 cm/sec. These figures are based on computations, but are in good agreement with direct measurements at one station in the passage. The transport amounts to about 20 million m³/sec. Similar conditions are encountered further north between the northern Riukiu Islands and the continental shelf, where, however, a weak countercurrent appears to be present on the left-hand side of the main flow. The maximum velocities in this profile are somewhat above 80 cm/sec, and the computed transport is 23 million m³/sec.

Off Shiono-misaki in latitude 33°N the character of the Kuroshio is closely related to the Florida Current to the south of Cape Hatteras. The transport of the Kuroshio between the coast and its outer limit is greatly increased, but the countercurrent on the right-hand side carries large amounts of water in the opposite direction. It is probable that a considerable annual variation takes place in the transport, but so far little information is available on the subject.

The temperatures of the Kuroshio water are comparable to those of the Florida Current, but the salinities are much lower, the maximum salinity in the Kuroshio being slightly less than 35.00°/₀₀, whereas the maximum salinity in the Florida Current is about 36.50°/₀₀. This difference reflects the general lower salinity of the Pacific as compared to the Atlantic. The temperature of the Kuroshio is subject to a large annual variation (p. 78) that is related to excessive cooling in winter by cold offshore winds. Off Shiono-misaki the annual range at the surface amounts to nearly 9°, and at 100 m the range is still nearly 4.5°.

THE KUROSHIO EXTENSION. In lat. 35°N, where the Kuroshio leaves the coast of Japan, it divides into two branches: one major branch that turns due east and retains its character as a well-defined flow as far as approximately long. 160°E, and one minor branch which continues toward the northeast as far as lat. 40°N, where it bends east. The major branch is evident on charts showing the anomaly of the surface temperature or the difference between air and surface temperatures in winter. According to Schott a tongue within which this difference is greater than 4°C extends east toward long. 170°E—that is, beyond the eastern limit of the well-defined flow. Between long. 155° and 160°E, considerable water masses turn toward the south and southwest, forming part of the Kuroshio Countercurrent, which runs at a distance of approximately 700 km from the coast as the eastern branch of a large whirl on the right-hand side of the Kuroshio.

According to the vertical sections of temperature and salinity in fig. 53 the Kuroshio Extension is, in long. 153°E, still a narrow current, but, continuing toward the east, a section in long. 162°E shows that the current has been broken up into a number of branches separated by eddies and countercurrents. The change corresponds to the one which, in the Atlantic Ocean, takes place between the regions to the south of the Grand Banks and to the north of the Azores (see fig. 41, p. 169).

The northern branch of the Kuroshio Extension becomes rapidly mixed with the cold waters of the Oyashio, which flow south close to the northeastern coasts of Japan, reaching nearly to 35°N. The extensive work of Japanese oceanographers shows that along the boundary between the Kuroshio Extension and the Oyashio numerous eddies develop within which a thorough mixing of the water masses takes place. From the sections in fig. 53, it is evident that to the north of lat. 36°N the Kuroshio

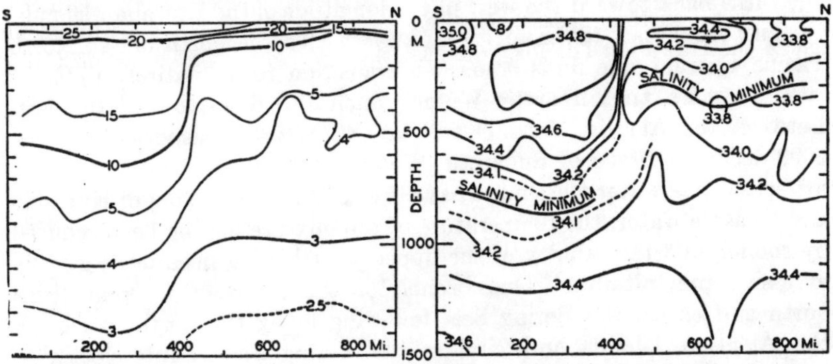

Fig. 53. Distribution of temperatures and salinities in a vertical section between lat. 30°28′N, long. 150°04′E, and lat. 42°40′N and long. 157°00′E (after Uda).

waters, which can be traced to lat. 41°N, have been greatly diluted by the Oyashio water. By these processes of mixing a water mass is gradually being formed which is present in the northwestern Pacific Ocean and which has been called the Subarctic Water of the North Pacific (fig. 40, p. 158).

These sections bring out the feature that to the north of lat. 36°N intermediate water of a salinity as low as 33.8°/₀₀ is found at a depth of 300 m or less. This water, which contains a considerable amount of oxygen, has probably been formed in winter at the convergence between the Kuroshio Extension and the Oyashio, and has sunk from the surface in a manner similar to the sinking of the Antarctic Intermediate Water. From the region of sinking, this intermediate water flows toward the east and spreads over the greater part of the North Pacific. To the south of lat. 36°N intermediate water of salinity between 34.0°/₀₀ and 34.1°/₀₀ is present at a depth of about 800 m, but this water represents the most northern extension of the intermediate water that flows north along the coast of Japan and turns around as part of the big whirl on the right-hand side of the Kuroshio.

The North Pacific Current. This term is applied to the general eastward flow of warm water to the east of long. 160°E. Details of this flow are not known, but from the *Bushnell* section of 1934, from the Aleutian to the Hawaiian Islands, it is evident that Kuroshio water crosses long. 170°W, because at a number of the *Bushnell* stations the temperature-salinity curves are similar to those of the Kuroshio water. The greater part of this water appears to turn around toward the south before reaching 150°W, and only a small portion continues and flows south between the Hawaiian Islands and the west coast of North America after having been mixed with waters of different origin. The main part of the North Pacific Current does not therefore extend across the Pacific Ocean, but turns back toward the west in the longitude of the Hawaiian Islands.

THE ALEUTIAN (SUBARCTIC) CURRENT. To the north of the North Pacific Current one finds a marked transition to an entirely different type of water, the Subarctic Water, which also flows toward the east. Between the Aleutian Islands and lat. 42°N the transport above the 2000-decibar surface of Subarctic Water amounts to about 15 million m^3/sec. This water mass must have been formed by mixing of Kuroshio and Oyashio water, the temperature of the mixture having been reduced by cooling and the salinity of the upper layer having been decreased by excessive precipitation. One branch of the Aleutian Current turns north and enters the Bering Sea, following along the northern side of the Aleutian Islands and circling the Bering Sea counterclockwise. In the Bering Sea, these waters are further cooled, and, flowing south, they reach the northern islands of Japan as the cold Oyashio. The amount of water in this gyral is not known, but since inflow and outflow from the Bering Sea must be nearly equal, it follows that the 15 million m^3/sec of Subarctic Water that continue east on the southern side of the Aleutian Islands are supplied by the Kuroshio, the waters of which have been completely transformed by mixing and external influences.

Before reaching the American coast the Aleutian Current divides, sending one branch toward the north into the Gulf of Alaska and another branch toward the south along the west coast of the United States. The former branch is part of the counterclockwise gyral in the Gulf of Alaska. It enters the Gulf along the American west coast, and, since it comes from the south, it has the character of a warm current in spite of the fact that it carries Subarctic Water. It therefore exercises an influence on climatic conditions similar, on a small scale, to that which the North Atlantic and the Norwegian Currents exercise on the climate of northwestern Europe. The major branch, which turns toward the south along the west coast of the United States, is known as the California Current, but before dealing with this it is desirable to discuss the warm-water currents between the Hawaiian Islands and the American West Coast.

THE EASTERN GYRAL IN THE NORTH PACIFIC OCEAN. The existence of an eastern gyral is recognized mainly by the character of the water masses and by the results of computations based on observations between the Hawaiian Islands and the coast of California and between the Hawaiian Islands and the Aleutian Islands. The study of water masses (fig. 40, p. 158) showed that a distinctly different water mass was present in the region to the east of the Hawaiian Islands, and that this mass was, in part, formed at the boundary between the warmer waters and the Subarctic Water between long. 130° and 150°W. Computation of currents and transport, the results of which are shown schematically in fig. 56, p. 205, leads to the conclusion that a clockwise-rotating gyral is present in the eastern North Pacific with its center to the northeast of the Hawaiian Islands. It is probable that the location of this gyral changes with the seasons and shifts from year to year, so that occasionally the gyral may lie entirely to the northeast of the Hawaiian Islands, whereas in other circumstances the Hawaiian Islands may lie inside the gyral. If the gyral is displaced considerably to the north, Equatorial Water may reach as far north as to the Hawaiian Islands, as was observed in 1939 at one of the *Bushnell* stations. At the surface the gyral is masked by wind-driven currents.

THE CALIFORNIA CURRENT. The California Current represents, as already explained, the continuation of the Aleutian Current of the North Pacific. The name is applied to the southward flow between lat. 48° and 23°N, where the Subarctic Water converges with the Equatorial (fig. 40, p. 158). The outer limit of the California Current is represented by the boundary region between the Subarctic Water and the Eastern North Pacific Water, and lies in lat. 32°N at a distance of approximately 700 km from the coast, according to the observations by the *Carnegie* in 1929, the *Louisville* in 1936, and the *Bushnell* in 1939. The total volume transport of the California Current above the 1500-decibar surface is probably not more than about 10 million m^3/sec, and, in view of the great width of the current, no high velocities are encountered except within local eddies. As a whole, the current represents a wide body of water that moves sluggishly toward the southeast.

In spring and early summer the California Current is a counterpart to the Peru Current, several characteristic features of the two currents being strikingly similar. During these months north-northwest winds prevail off the coast of California, giving rise to upwelling that begins mostly in March and continues more or less uninterruptedly until July. Records of surface temperatures show that on the coast the lowest temperatures regularly occur in certain localities that are separated by regions with higher surface temperatures. This is demonstrated by fig. 54, in which are plotted the surface temperatures between latitude 32°N and 45°N along the American west coast in March to June and in November, December, and January. In the regions of intense upwelling the spring temperatures are lower than the winter temperatures, but in regions of less intense upwelling they are higher.

Recent work of the Scripps Institution of Oceanography demonstrates that from the areas of intense upwelling tongues of water of low temperature extend in a southerly direction away from the coast, and that these tongues are separated from each other by tongues of higher temperature extending in toward the coast. Within the tongues of higher temperature the flow is directed to the north, whereas within the tongues of low temperature the flow is directed to the south. Thus, a series of swirls appear on the coastal side of the current similar to those that Gunther has demonstrated within the Peru Current.

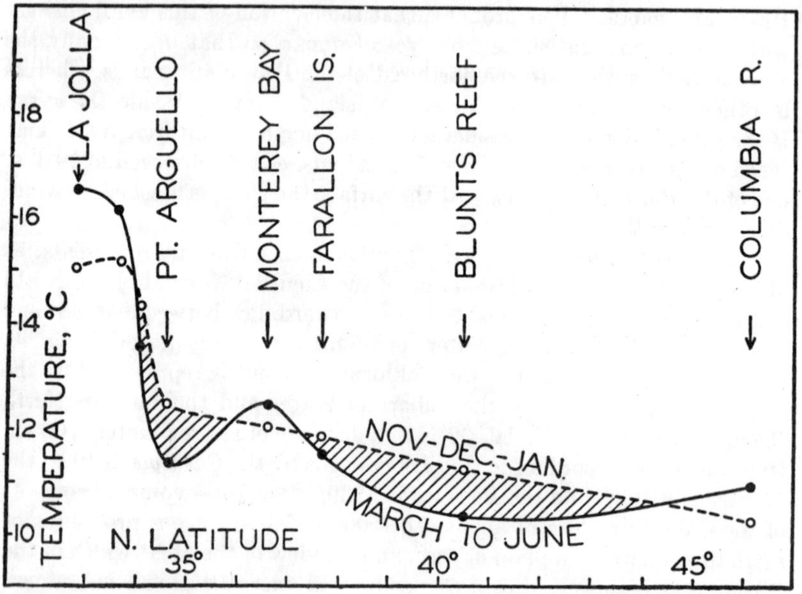

Fig. 54. Surface temperatures along the coast of California in March to June and in November to January. In regions of intense upwelling the average surface temperature is lower in March to June than in December to January.

To the north of latitude 30°N the two most conspicuous centers of upwelling are located in latitudes 35°N and 41°N. The two swirls associated with these centers of upwelling and the alternating movements away from and toward the coast are shown in fig. 55, which is based on observations in May to July, 1939. The coast to the south of 34°N is under the influence of water that returns after having traveled a long distance as surface water, and, correspondingly, the surface temperatures are much higher there than those encountered where the water was recently drawn to the surface (fig. 54). A third region of intense upwelling is found perhaps in about latitude 24°N, on the coast of Lower California.

The upwelling water rises from only moderate depths, probably less than 200 m, and the phenomenon therefore represents only an overturning of the upper layers such as is the case in other regions that have

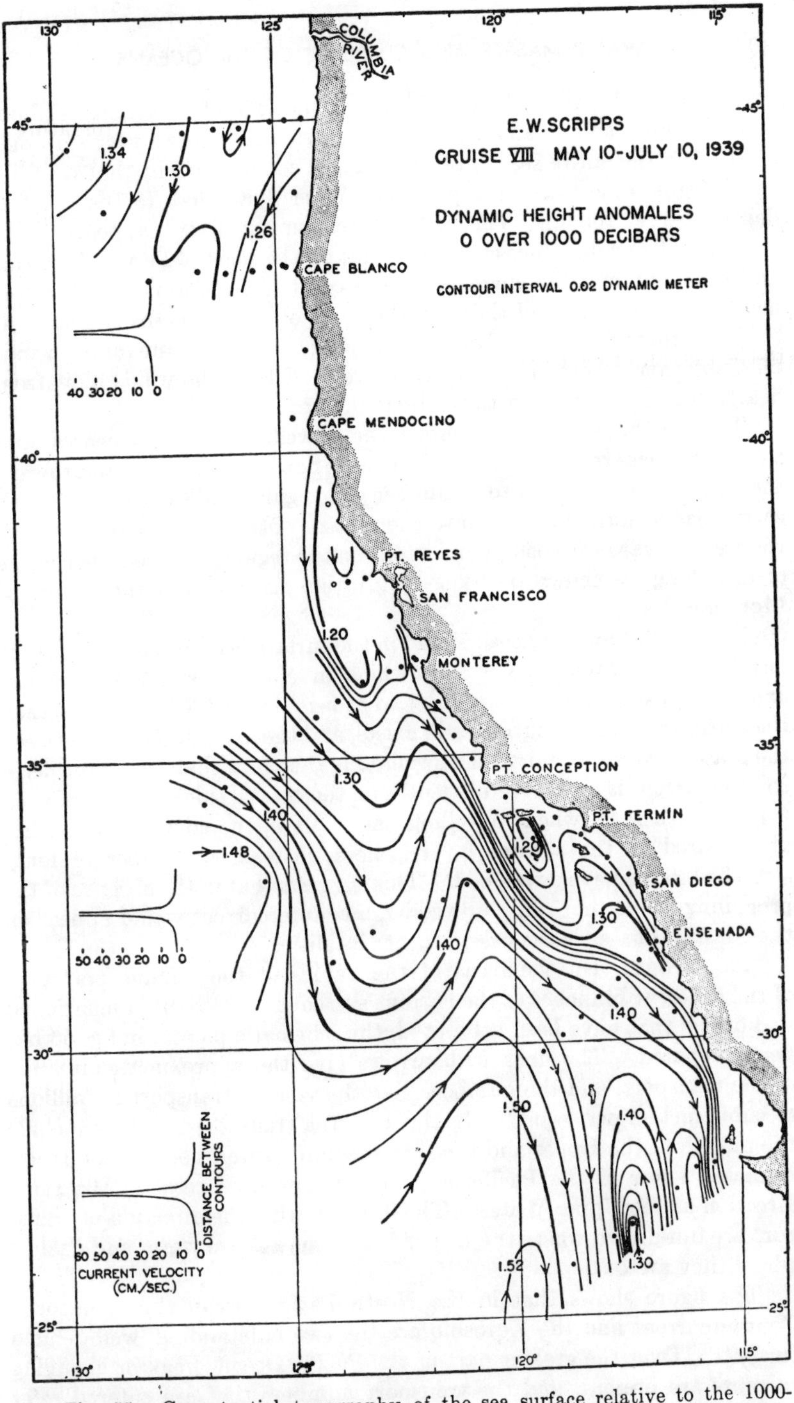

Fig. 55. Geopotential topography of the sea surface relative to the 1000-decibar surface off the American west coast in May to July, 1939, showing flow alternating away from and toward the coast.

been examined. According to McEwen the rate of upwelling is about 20 m a month, and on an average the process is therefore a very slow one.

During the entire season of upwelling a countercurrent that contains considerable quantities of Equatorial Water flows close to the coast at depths below 200 m. This subsurface countercurrent appears to be analogous to a subsurface countercurrent off the coast of Peru. In spring and early summer the currents off the coast of California are therefore nearly a mirror image of those off the coast of Peru, but the similarity is found in this season only because the character of the prevailing winds off California changes in summer, whereas off Peru the winds blow from nearly the same direction throughout the year.

Toward the end of the summer the upwelling gradually ceases, and the more or less regular pattern of currents flowing away from and toward the coast breaks down into a number of irregular eddies, some of which carry coastal waters far out into the ocean. Other eddies carry oceanic waters in toward the coast, particularly in the regions between the centers of upwelling, as shown by Skogsberg in his discussion of the waters of Monterey Bay.

In the fall, upwelling ceases, and in the surface layers a countercurrent develops, the Davidson Current, which in November, December, and January runs north along the coast to at least lat. 48°N. In this season the subsurface countercurrent still exists, and the main difference between the seasons without and with upwelling is therefore that in the former a countercurrent is present at all depths on the coastal side of the California Current, whereas, when upwelling takes place, the countercurrent has disappeared in the surface layer, where, instead, a number of long-stretched swirls have developed. This suggests that in the absence of the prevailing winds that cause upwelling, a countercurrent would appear on the coastal side, as is the case in other localities.

Transport. On the basis of the values of the volume transport of the different branches of the current system that have been mentioned and others that have been computed, the schematic picture in fig. 56 has been prepared. The lines with arrows give the approximate direction of the transport, and the numbers give the volume transport in millions of cubic meters per second. In this case the transport numbers include the motion of the Upper and the Intermediate Water, because the Intermediate Water of the Pacific appears to flow in general in the same direction as the Upper Water. The lines showing the direction of transport are full-drawn where the Upper Water masses are warm and dashed where they are cold.

The figure shows that in the North Pacific Ocean the Equatorial Countercurrent and the Kuroshio are the two outstanding, well-defined currents. Over the greater part of the Pacific Ocean, weak or changing currents are present, and the transport numbers that are entered refer therefore to broad cross sections. The figure is intended to bring out only the major features that have been discussed, and represents the first attempt at a synthesis of the available information as to the circulation in the North Pacific Ocean.

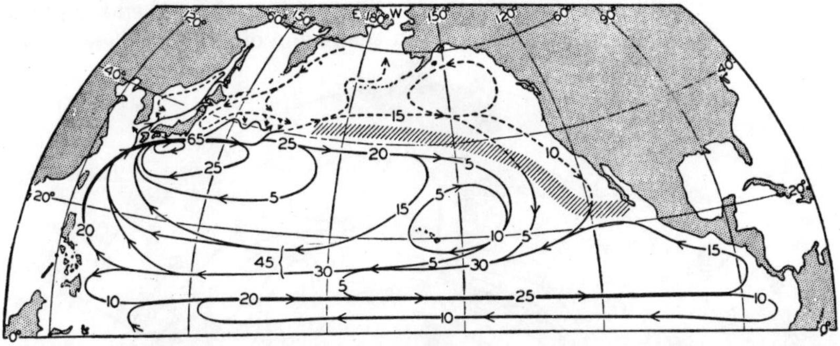

Fig. 56. Transport chart of the North Pacific. The lines with arrows indicate the approximate direction of the transport above 1500 m, and the inserted numbers indicate the transported volumes in millions of cubic meters per second. Dashed lines show cold currents, full-drawn lines show warm currents.

Currents of the Antarctic Ocean

The Antarctic Circumpolar Current runs from west to east around the Antarctic Continent, but is locally deflected from its course, partly by the distribution of land and sea and partly by the submarine topography. The effect of the submarine topography is seen in fig. 57, which shows the total volume transport relative to the 3000-decibar surface as computed by means of *Discovery* data, in addition to the results from a number of *Meteor* stations in the South Atlantic. In the figure the volume transport is shown by lines that are approximately in the direction of transport and which have been drawn with such intervals that the transport between two lines equals 20×10^6 m^3/sec. As the current passes around the Antarctic Continent, it is seen that when approaching a submarine ridge the current bends to the left, and after passing the ridge it bends to the right. Some of these bends appear on charts of the surface currents (see chart 4).

Besides the bends that are associated with the bottom topography, the effects of the distribution of land and sea and of the currents in the adjacent oceans are also evident. The location of the Drake Passage naturally forces the current further south than in any other region, but on the eastern side of South America a branch of the current, the Falkland Current, turns north. Along the east coast of South Africa the Agulhas Stream flows south, partly turning into the Atlantic Ocean, but mainly bending around toward the east, and the narrowness of the Antarctic Current in longitude 30°E is probably related to the effect of the Agulhas

Stream. When the Circumpolar Current has passed New Zealand, a bend toward the north, which corresponds on a small scale to the Falkland Current, is evident.

The transport line marked "zero" is not continuous around the Antarctic Continent, but begins in the Weddell Sea area and ends against the coast of the Antarctic Continent in the region of the Ross Sea. This

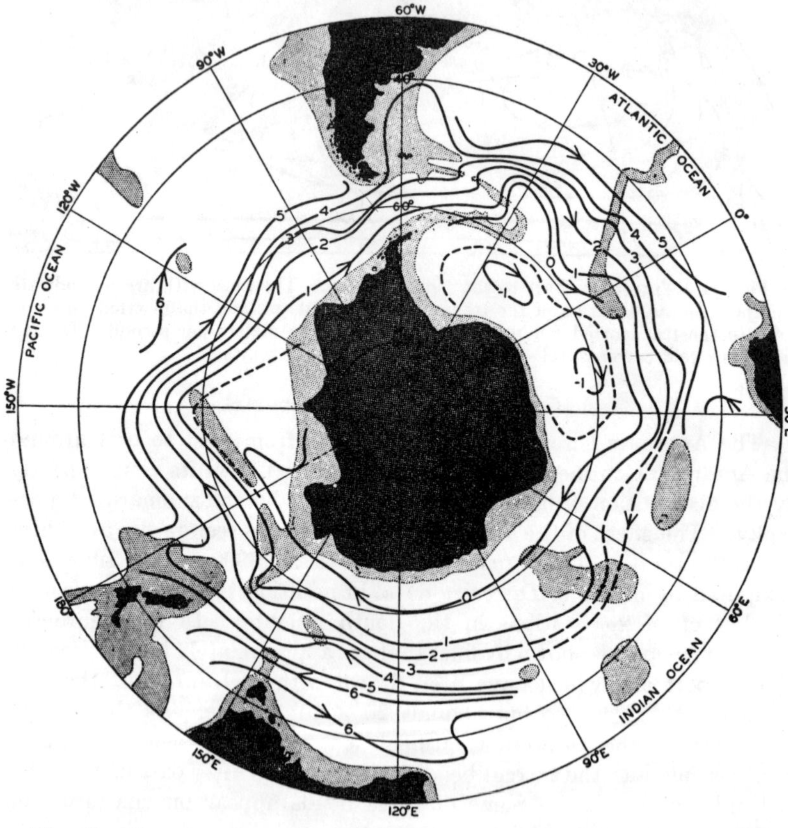

Fig. 57. Transport lines around the Antarctic Continent. Between two lines the transport relative to the 3000-decibar surface is about 20 million m³/sec.

feature indicates that our representation is not entirely correct and that the transport does not exactly take place along the lines that have been entered, because actual lines of equal transport must be closed lines and cannot end against the shores. Some transport must take place across the lines, and the fact that our "zero" line stops against the continent to the west of Drake Passage indicates that in this narrow passage a certain amount of piling up of water takes place and that transport occurs here across the contour lines. If the flow is referred to the dynamic topography of the isobaric surfaces, this means that in Drake Passage and the Scotia Sea the water does not flow parallel to the dynamic contour lines but flows uphill. Within the Antarctic Ocean, this region is the only one in which there is evidence of a marked discrepancy between the lines of equal transport and the actual flow of water, and on the whole it may be assumed that the transport lines give a fairly correct picture of the character of the Circumpolar Current.

A flow to the west near the Antarctic Continent is evident only in the Weddell Sea area, where an extensive cyclonic motion occurs to the south of the Circumpolar Current. The water masses that take part in this cyclonic movement, however, are small compared to those of the Circumpolar Current, and their velocities are small.

Within the subantarctic region the current is also directed in general from west to east, but only the southern portion of the waters close to the Antarctic Convergence circulates around the Antarctic Continent and forms part of the Circumpolar Current. In the Pacific Ocean the northern portion belongs to the current system of that ocean. Thus, a strict northern limit of the Antarctic Ocean cannot be established on oceanographic principles, but a boundary region has to be considered that, depending upon the point of view, may be assigned either to the Antarctic Ocean or to the adjacent oceans.

The total transport of the Antarctic Circumpolar Current must be greater than is apparent from fig. 57, which shows only the transport relative to the 3000-decibar surface. According to the figure the transport through Drake Passage is about 90 million m^3/sec, whereas Clowes' computations gave 110 millions above the 3500-decibar surface. It is therefore probable that the absolute transport between the lines in fig. 57 is not 20 million, but at least 25 million m^3.

This discussion of the Circumpolar Current is based entirely on the distribution of mass as derived from deep-sea oceanographic observations. The surface currents have been determined independently by means of ships' records and show that the flow of the surface water is governed partly by the distribution of mass and partly by the effect of the prevailing winds. Near the Antarctic Continent, easterly and southeasterly winds blow away from the large land masses, but between latitudes 60° and 40°S strong westerly winds prevail. Correspondingly one finds, as seen on chart 4, westward surface currents prevailing near the border of the Antarctic Continent and eastward surface currents at some distance from the coast. In the Southern Hemisphere the wind drift deviates to the left of the direction of the wind, and consequently the eastward surface current shows a component toward the north. A divergence must be present between the westward and the eastward surface currents, drawing deep water toward the surface, and in the sections (fig. 58) this divergence shows up close to the continent by the high temperature and high salinity of the water at a depth of 100 m. Between the Antarctic and Subtropical Convergences the surface current is generally directed toward the east, but on the northern side of the Antarctic Convergence

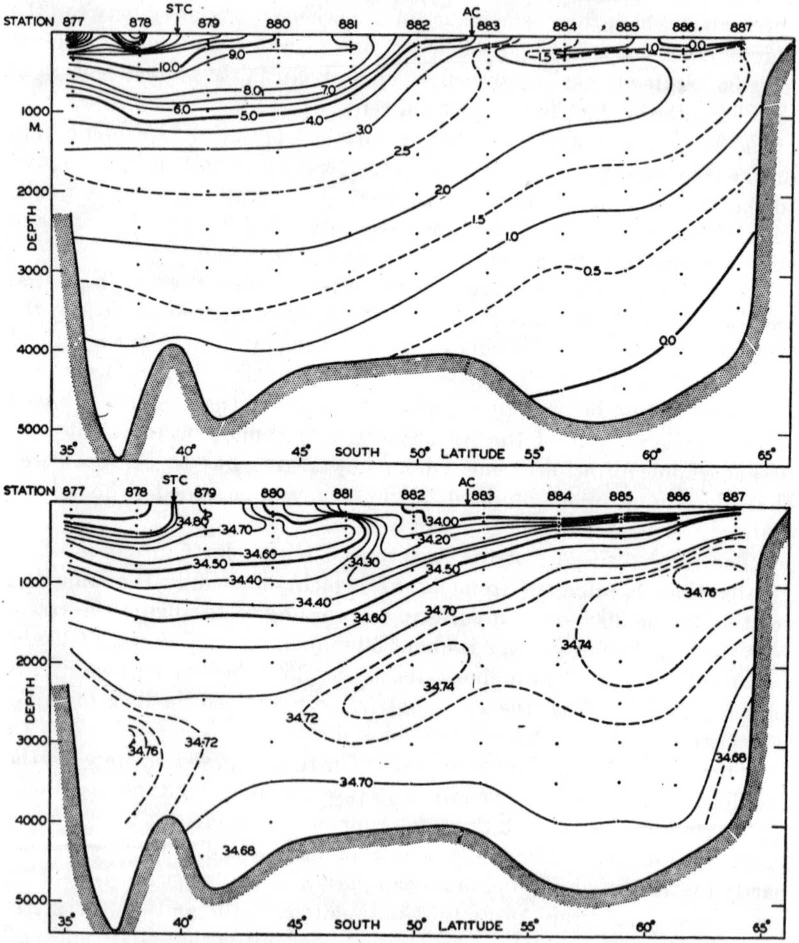

Fig. 58. Distribution of temperature and salinity in a vertical section from Cape Leeuwin, Australia, to the Antarctic Continent (after Deacon).

it shows a component to the south, and at the southern boundary of the Subtropical Convergence, a component to the north. The component to the south must be associated with the development of the Antarctic Convergence, which has not yet been satisfactorily explained.

As a consequence of the divergence and convergence at the surface a transverse circulation that is superimposed upon the general flow from west to east must be present. Such a transverse circulation is evident from the vertical sections in fig. 58, in which, between latitudes 45° and 63°S, the deep water seems to climb from a depth of about 3000 m to within 200 m of the surface. The deep water does not appear to reach the very surface, but within the entire antarctic region a considerable amount of deep water of high salinity must be added to the surface layer, because, otherwise, the salinity would be lowered by the excess of precipitation and the addition of fresh water by the melting of antarctic icebergs. A southward flow of deep water is also required for balancing the transport to the north of the surface water by the prevailing winds. In the section shown in fig. 58, sinking of water near the Antarctic Continent probably does not take place, but in the Weddell Sea this sinking is of great importance, and a movement of the deep water toward the continent is there necessary for the formation of the bottom water. Part of this bottom water flows away from the Antarctic Continent, but part mixes with the deep water and returns with this water to the antarctic regions (p. 214).

At the Antarctic Convergence, water of relatively low salinity and low temperature sinks. A small portion of the sinking water appears in some areas to return toward the south at a depth of a few hundred meters, but the greater part continues toward the north, forming the tongues of Antarctic Intermediate Water that in all oceans can be traced to great distances from the antarctic region. It is probable that this water gradually mixes with the underlying deep water and returns to the Antarctic with the deep water. Within the subantarctic regions the upper water appears to flow toward the south, but the nature of this flow is not fully understood.

The Antarctic Water that sinks at the Antarctic Convergence has a temperature of 2.2° and a salinity of 33.80°/₀₀. When sinking it is rapidly mixed with surrounding waters, and a *water mass*, the Antarctic Intermediate Water, is formed that spreads mainly toward the north, being characterized at its core by a salinity minimum. Owing to continued mixing the characteristic temperature-salinity relation of the intermediate water changes as the distance from the Convergence increases, and, since layers of gradual transition are found both above and below the water mass, no distinct boundaries can be introduced.

On the basis of these considerations, one arrives at the schematic picture of the transverse circulation that is shown in the block diagram in fig. 59. In the diagram the formation of bottom water has been taken into account, and, furthermore, it is indicated that the deep-water flow toward the Antarctic Continent is strengthened by addition of both intermediate and bottom water. The main features of the transverse circulation are represented by the sinking of cold water near the Antarctic Continent, the climb of the deep water toward the surface, the northward flow of surface water, and the sinking of the Antarctic Intermediate Water.

From the foregoing it is evident that one cannot consider the Antarctic Circumpolar Water Mass as a body of water that circulates around and around the Antarctic Continent without renewal. On the contrary, one has to bear in mind that water from the antarctic region is carried toward the north and out of the region both near the surface and near the bottom, and that deep water from lower latitudes is drawn into the

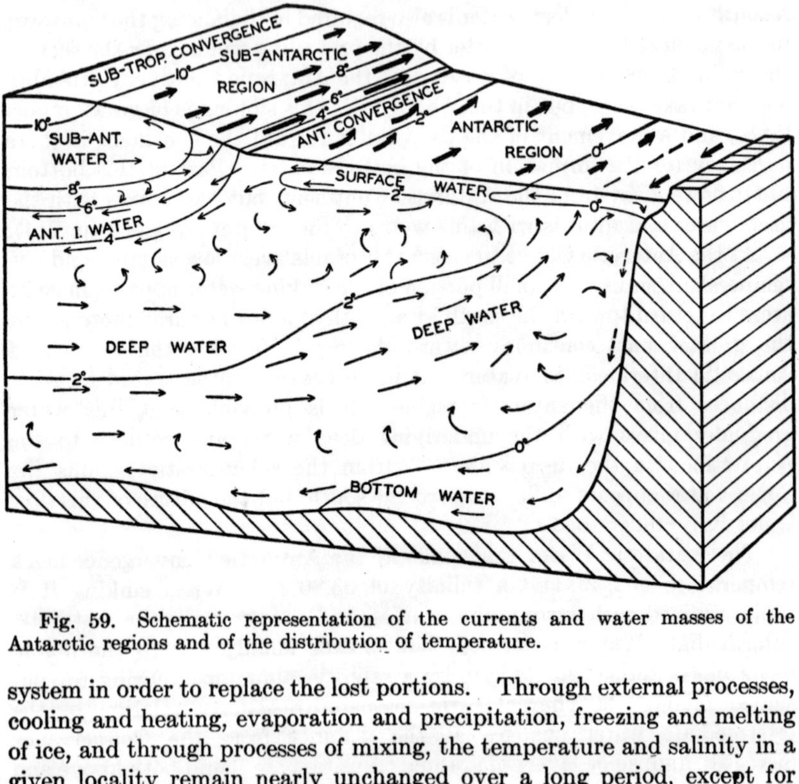

Fig. 59. Schematic representation of the currents and water masses of the Antarctic regions and of the distribution of temperature.

system in order to replace the lost portions. Through external processes, cooling and heating, evaporation and precipitation, freezing and melting of ice, and through processes of mixing, the temperature and salinity in a given locality remain nearly unchanged over a long period, except for such changes as are associated with displacements of the currents. Thus, the stationary distribution of conditions that characterizes the entire Antarctic Ocean represents a delicate balance between a number of factors which tend to alter the conditions. On the other hand, an individual water particle, which describes a most complicated path, is subjected to great changes. It may lose its identity by mixing with adjacent water masses, or it may have its temperature and salinity radically altered if brought to the surface.

When examining the dynamics of the Circumpolar Current, frictional forces have to be taken into consideration. The stress that the wind exerts at the sea surface cannot be balanced by a piling-up of water in the direction of flow (p. 121), because the current is continuous around the earth, and therefore it must be balanced by other frictional stresses. The velocities along the bottom are too small to give rise to bottom friction, and one is therefore led to the conclusion that great stresses due to lateral mixing exist at the boundaries of the current. The torque exerted by these stresses along the quasi-vertical boundary surfaces of the current must balance the torque exerted by the wind on the surface, but so far no attempt has been made at an analysis of the dynamics of the current, taking friction into account.

The Deep-Water Circulation of the Oceans

The manner in which the deep and bottom water is formed has been discussed (p. 155) and certain statements as to the deep-water circulation have been made, but so far no general review of the deep-water circulation has been presented. Table 22 has been prepared in order to facilitate such a review. It contains temperatures, salinities, and oxygen values of the deep and bottom water at fifteen selected stations, six in the Atlantic, three in the Indian, and six in the Pacific Ocean, including the adjacent parts of the Antarctic Ocean. The oxygen content of the water has not been discussed, but is included here because it is useful when examining the slow spreading of deep water. The content is generally high near the sea surface, where oxygen is absorbed from the air, but low in "old" deep and bottom water, in which oxygen has been used up by respiration of marine organisms or for decomposition of organic remains. The content of table 22 will not be dealt with separately but must be examined as the discussion proceeds.

In order to understand the deep-water circulation, one has to bear in mind that deep and bottom waters represent water whose density became greatly increased when the water was in contact with the atmosphere, and that this water, by sinking and subsequent spreading, fills all deeper portions of the oceans. The most conspicuous formation of water of high density takes place in the subarctic and in the antarctic regions of the Atlantic Ocean. The deep and bottom water in all oceans is derived mainly from these two sources, but is to some extent modified by addition of high-salinity water flowing out across the sills of basins in lower latitudes, particularly from the Mediterranean and the Red Sea.

In the North Atlantic Ocean, North Atlantic Deep and Bottom Waters flow to the south, the flow being reinforced, and the upper deep water being modified, by the high-salinity water flowing out through the Strait of Gibraltar. The newly formed deep and bottom water has a high oxygen content that decreases in the direction of flow. Fig. 60 and the values given in table 22 demonstrate the character of these waters and show particularly the increase in salinity at moderate depth that is due to addition of Mediterranean water. Antarctic Bottom Water flows in

TABLE 22

TEMPERATURE, SALINITY, AND OXYGEN CONTENT BELOW 2000 METERS AT SELECTED STATIONS

(Met = *Meteor*, Atl = *Atlantis*, Da = *Dana*, B.A.E. = B.A.N.Z. Antarctic Research Exped., Di = *Discovery*, B = *Bushnell*, EWS = *E. W. Scripps*)

Atlantic Ocean, and Atlantic Antarctic Ocean

Depth (m)	Met 127			Atl 1223					
	March 12, 1935 50°27.5′N, 40°14.5′W			April 19, 1932 33°19′N, 68°18′W			March 17, 1927 19°16′N, 27°27′W		
	°C	S°/oo	O_2	°C	S°/oo	O_2	°C	S°/oo	O_2
2000..........	3.32	34.92	6.30	3.62	34.97	6.08	3.54	34.98	5.07
2500..........	.22	.93	.26	.37	.97	.04	.11	.96	.30
3000..........	2.97	.93	.17	2.95	.96	5.99	2.76	.94	.27
3500..........	.63	.95	.28	.61	.94	6.03	.49	.92	.32
4000..........	.38	.95	.34	.45	.92	.06	.39	.89	.42
5000..........	—	—	—	.54	.90	5.88	—	—	—

Depth (m)	Met 86			Met 135			Met 129		
	Dec. 4, 1925 32°49′S, 40°01′W			March 7, 1926 39°46′S, 22°12′E			Feb. 22–23, 1926 58°53′S, 4°54′E		
2000..........	2.97	34.77	4.71	2.68	34.76	4.70	−0.26	34.67	3.37
2500..........	3.10	.90	5.53	.54	.82	.99	−0.36	.67	.52
3000..........	2.86	.92	.65	.32	.81	5.14	−0.42	.66	.59
3500..........	.15	.89	.46	1.93	.80	.04	−0.51	.65	.67
4000..........	0.77	.69	4.88	.42	.78	4.97	−0.55	.64	.79

Indian Ocean, and Indian Antarctic Ocean

Depth (m)	Da 3917			B.A.E. 75			Di 858		
	Dec. 5, 1929 1°45′N, 71°05′E			March 19, 1930 36°41′S, 114°55′E			April 24, 1932 60°10′S, 63°55′E		
	°C	S°/oo	O_2	°C	S°/oo	O_2	°C	S°/oo	O_2
2000..........	2.68	34.79	—	2.70	34.60	—	0.90	34.72	4.41
2500..........	.09	.76	3.22	.28	.68	—	.63	.70	.51
3000..........	1.84	.79	2.78	1.93	.72	—	.34	.68	.48
3500..........	.66	.75	3.17	.59	.73	—	.11	.68	.66
4000..........	.71	.74	.61	.25	.74	—	−0.09	.67	.77

Pacific Ocean, and Antarctic Pacific Ocean

Depth (m)	B			EWS VIII-77			Da 3745		
	Aug. 18, 1934 50°30′N, 175°16′W			July 3, 1939 28°02′N, 122°08′W			July 8, 1929 3°18′N, 129°02′E		
	°C	S°/oo	O_2	°C	S°/oo	O_2	°C	S°/oo	O_2
2000..........	1.88	34.58	1.64	2.13	34.61	1.89	2.24	34.67	2.56
2500..........	.72	.59	2.20	1.83	.63	2.44	1.86	.69	.82
3000..........	.61	.64	.58	.65	.64	.72	.65	.69	3.15
3500..........	.50	.68	3.00	.55	.67	3.00	.62	.70	.26
4000..........	—	—	—	—	—	—	.57	.70	.27

Depth (m)	Da 3561			Da 3628			Di 950		
	Sept. 24, 1928 4°20′S, 116°46′W			Dec. 15, 1938 31°25′S, 176°25′W			Sept. 7, 1932 59°05′S, 163°46′W		
	°C	S°/oo	O_2	°C	S°/oo	O_2	°C	S°/oo	O_2
2000..........	2.30	34.63	2.53	2.42	34.60	3.32	1.70	34.73	4.27
2500..........	.01	.64	.75	.16	.64	.25	.34	.73	.33
3000..........	1.84	.65	.86	1.89	.68	.75	.09	.72	.37
3500..........	.70	.66	.96	.49	.72	4.27	0.91	.71	.20
4000..........	.64	.67	3.00	.22	.72	.52	.87	.70	.06
5000..........	—	—	—	.02	.71	.55	—	—	—

the opposite direction, from south to north, and has been traced beyond the Equator to latitude 35°N.

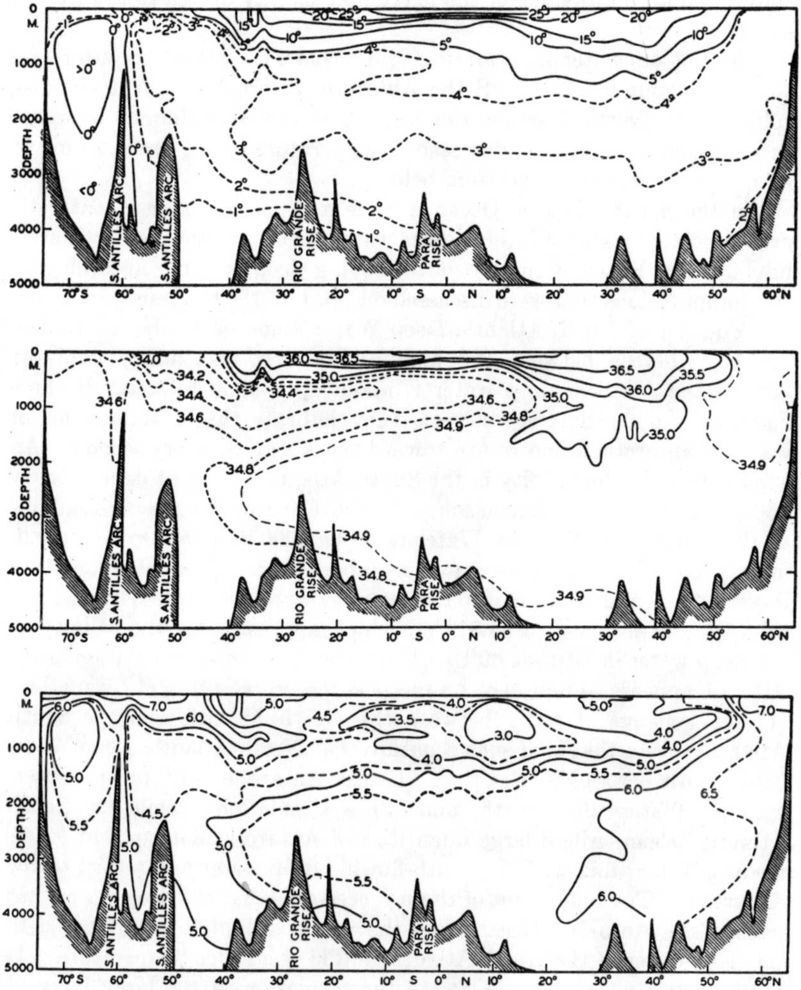

Fig. 60. Vertical sections showing distribution of temperature, salinity, and oxygen in the Western Atlantic Ocean (after Wüst).

The North Atlantic Deep Water crosses the Equator and continues toward the south above the Antarctic Bottom Water. On the other hand, it sinks below the Antarctic Intermediate Water and therefore, in the South Atlantic Ocean, it becomes "sandwiched" between the Antarctic Intermediate Water and the Antarctic Bottom Water, both of which are of lower salinity. In a vertical section the deep water of the South Atlantic Ocean is therefore characterized by a salinity maximum, but, owing to mixing with the overlying and the underlying water, the absolute value of the maximum salinity decreases toward the south.

The bottom water of antarctic origin is colder than the deep water, and south of about latitude 20°S the Antarctic Intermediate Water is also colder. In a vertical section the deep water therefore shows a temperature maximum and also a decreasing temperature to the south, owing to admixture from above and from below.

In the South Atlantic Ocean a large amount of water of antarctic origin, bottom water or intermediate water, returns to the Antarctic after having been mixed with the south-moving deep water. According to the computations that were discussed on pp. 116, 186 the transport across the Equator of North Atlantic Deep Water amounts to about 9 million m^3/sec, whereas between 20° and 30°S the corresponding transport toward the south of deep water is about 18 million m^3/sec. If these figures are approximately correct, they indicate that 9 million m^3 of water of antarctic origin return toward the Antarctic every second. An examination of the salinity in the South Atlantic Ocean at depths below 1600 m confirms this conclusion. At the Equator the average salinity of the North Atlantic Deep Water between 2000 and 3500 m is approximately 34.93°/₀₀. If this water is mixed with an equal amount of intermediate and bottom water of salinity approximately 34.7°/₀₀, the salinity will decrease to 34.81°/₀₀, which approximates the salinity of the deep water in latitude 40°S. Thus, the deep-water circulation of the Atlantic appears to represent a superposition of two types of circulation: (1) an exchange of water between the North Atlantic and the South Atlantic Ocean that is of such a nature that North Atlantic Deep Water flows south across the Equator, whereas Antarctic Bottom and Intermediate Waters flow north, and (2) a circulation within the South Atlantic Ocean, where large quantities of Antarctic Bottom and Intermediate Water mix with the south-flowing deep water and return to the Antarctic. The final result of these processes is that the deep water that reaches the Antarctic Ocean from the north is diluted as compared to the deep water of the North Atlantic and is of a lower temperature. It is this water which contributes to the formation of the large body of Antarctic Circumpolar Water flowing around the Antarctic Continent. The circulation that has been described is present in the western part of the South Atlantic Ocean, but in the eastern part it is impeded by the Walfish ridge.

In the Indian Ocean there is no large southward transport of deep water across the Equator. The data in table 22 show that to the north of the Equator the deep water contains an admixture of Red Sea Water, which maintains a relatively high salinity down to depths exceeding 3000 meters, but to the south of the Equator the temperature-salinity values indicate only a slight effect of the Red Sea Water. In the southern part of the Indian Ocean an independent circulation must be present. The deep water from the South Atlantic Ocean continues into the Indian Ocean and is particularly conspicuous in the western part, where maximum salinities of 34.80°/₀₀ have been observed. This water flows mainly toward the east, being somewhat diluted by admixtures of intermediate and bottom waters. On the other hand, Antarctic Intermediate Water flows north, and the bottom temperatures demonstrate that bottom water also moves north, and these water masses must return again to the south. It is probable that the intermediate water and the bottom water mix with the deep water and that the return flow takes place within the latter. A slight admixture of deep water from the region to the north of the Equator that is compensated for by bottom water penetrating across the Equator appears to maintain the salinity of the Indian Ocean Deep Water at a higher level than would be the case if no addition of saline water took place. Thus the influence of the Red Sea can probably be traced to the Antarctic.

From the Indian Ocean the Antarctic Circumpolar Water with its components of Atlantic and Indian Ocean origin enters the Pacific Ocean. The *Discovery* and *Dana* observations in the Tasman Sea between Australia and New Zealand, and in the Pacific to the east of New Zealand, show that the salinity of the deep and bottom water has been reduced so much that the maximum values lie between 34.72 and 34.74°/₀₀. These maximum salinities are found at depths between 2500 and 4000 m, the salinity of the water close to the bottom being slightly lower. From the region where the deep water enters the Pacific Ocean the salinity decreases both toward the north and toward the east. The *Discovery* data indicate that below the Antarctic Convergence a core of water of salinity higher than 34.72°/₀₀ is found which represents water of the Circumpolar Current, but to the north of this region values below 34.7°/₀₀ prevail, increasing uniformly toward the bottom. The *Carnegie* and the *Dana* data similarly show that north of 40°S the highest salinities are found near the bottom, and therefore the structure of the water masses of the Pacific differs completely from that found in the other oceans, where the highest salinities are encountered in the deep water and not in the bottom water. This feature can be explained if one assumes that in the South Pacific Ocean there also exists a circulation which is similar to that of the South Atlantic and Indian Oceans—namely, that intermediate and bottom water flow to the north and that a flow of deep water to the south takes place. This north-south circulation is superimposed upon a general flow from west to east. The Pacific Deep Water is, therefore, of Atlantic and Indian origin, but has become so much diluted by admixture of intermediate and bottom water that the salinity maximum has disappeared.

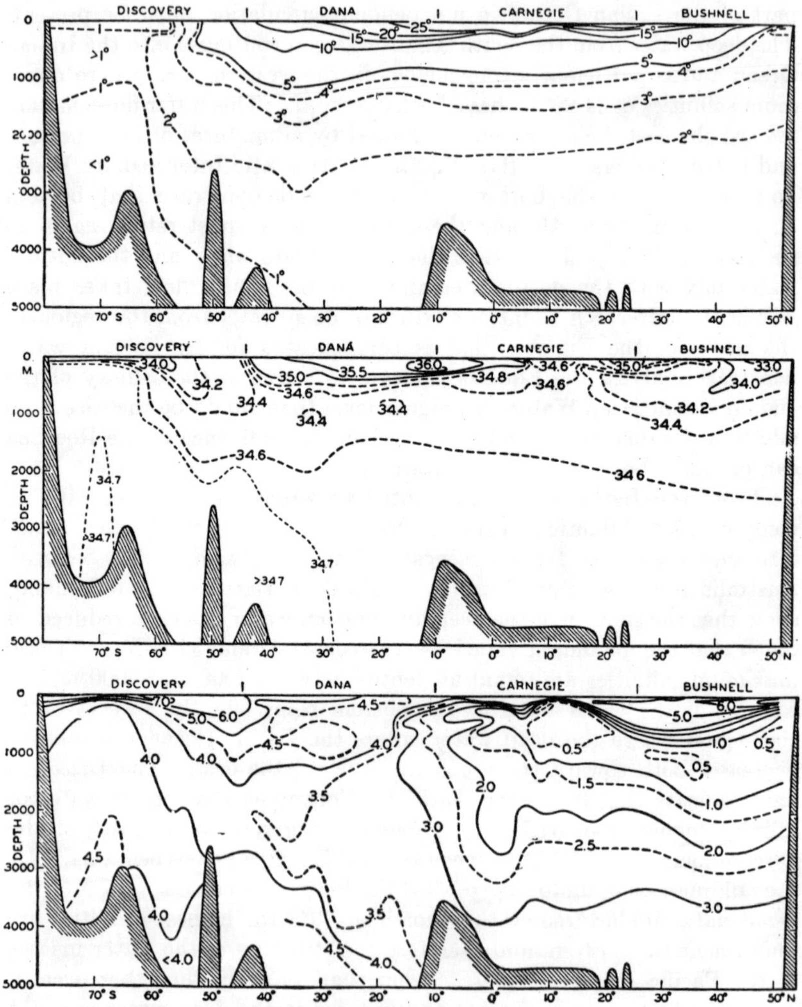

Fig. 61. Vertical sections showing distribution of temperature, salinity, and oxygen in the Pacific Ocean, approximately along the meridian of 170°W.

The more or less closed systems of the deep-water circulation that are present in the Southern Hemisphere between the Antarctic Ocean and the Equator are related to the fact that near the Equator all ocean currents tend to flow in east-west directions and that transport of water across the Equator takes place only when required to maintain the same sea level in both hemispheres. That is the case in the Atlantic Ocean, where deep water is formed in the Northern Hemisphere. The water that sinks in high latitudes or flows out of the Mediterranean Sea spreads at great depths, continues across the Equator, and is replaced by horizontal flow from the South Atlantic Ocean. In the Indian Ocean a similar mechanism operates, but on a very small scale, because sinking takes place only in the Red Sea, and the amounts of deep water that are formed there are small and exercise a significant influence only upon conditions to the north of the Equator. In the Pacific Ocean no mechanism exists that will give rise to a considerable exchange of water across the Equator, because conditions are such that no deep water is formed in the North Pacific.

On the other hand, since continuity must exist between the South and the North Pacific, the depths of the North Pacific Ocean are filled by water of the same character as that which is found in the northern portion of the South Pacific. The most accurate data that are available indicate that a very small exchange of deep water takes place between the two hemispheres and that a sluggish motion to the north may occur on the western side of the Pacific Ocean, whereas a sluggish motion to the south may occur on the eastern side. The uniform character of the Pacific Deep Water is illustrated in fig. 61.

In the preceding discussion the term "flow of water" has been used freely, but one has to deal actually with such slow and sluggish motion that the term "flow" is not appropriate, since the average velocities must often be measured in fractions of a centimeter per second.

In conclusion it can be stated that appreciable exchange of deep and bottom water across the Equator takes place in the Atlantic Ocean only. It is rudimentarily present in the Indian Ocean and practically absent in the Pacific. Superimposed upon such an exchange between the hemispheres there exist independent circulations in the three southern oceans, because Antarctic Intermediate and Bottom Waters return to the Antarctic as deep water. This circulation is well established in the Atlantic Ocean, and in the Indian and Pacific Oceans the existence of such a circulation is derived partly by analogy with the Atlantic Ocean and partly by an examination of the few available precise salinity observations.

Ice in the Sea

Ice and Icebergs in the Antarctic. Two forms of ice are encountered in the Antarctic Ocean: *sea ice,* which is formed by freezing of sea water, and *icebergs,* which represent broken-off pieces of glaciers. The appearance of both sea ice and icebergs varies widely. Several classifications have been proposed by Transehe, Smith, Maurstad, and Zukriegel, and several codes for reporting ice have been introduced, one of which (Maurstad's) has been adopted by the International Meteorological Organization. The classification of sea ice deals with forms produced under different conditions of freezing, forms that are due to the tearing apart and packing together of the ice under the action of winds and tides and forms that result from disintegration and melting

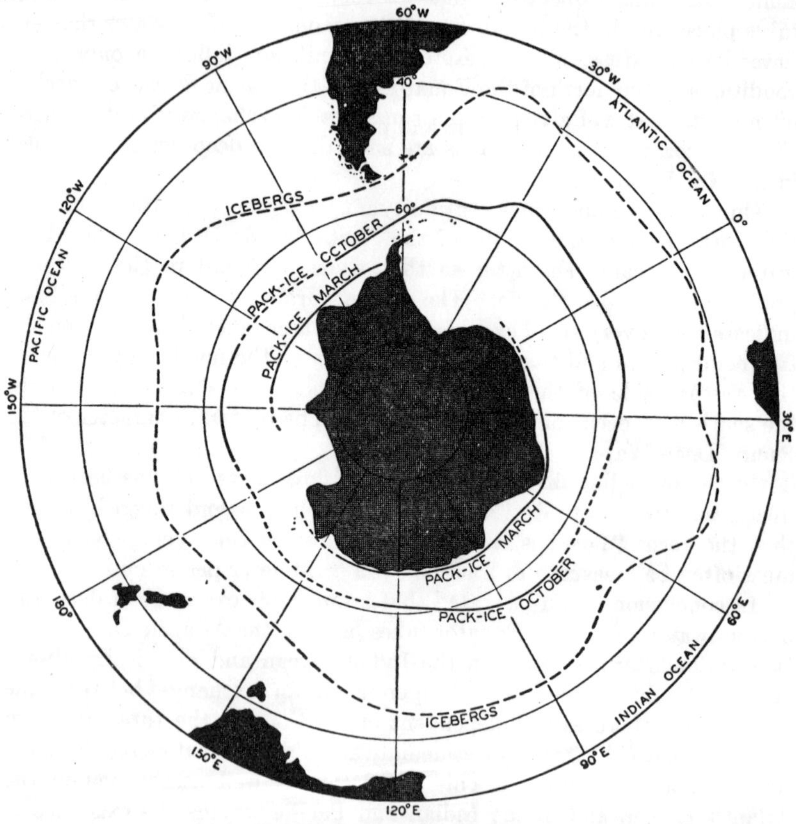

Fig. 62. Average northern limits of pack ice around the Antarctic Continent in March and October (after Mackintosh and Herdman) and average northern limit of icebergs according to British Admiralty Chart No. 1241.

of the ice. The classification of icebergs is based on the shape of the bergs, which depends partly upon the type of glacier from which they originate and partly upon the processes of melting and destruction to which they have been subjected.

The great mass of sea ice around the Antarctic Continent is popularly described by the somewhat loose term "pack ice," by which is meant drifting fields of close ice, consisting mainly of relatively flat ice floes, the edges of which are broken and hummocked. The pack ice is kept in motion by winds and currents. The effect of the wind is conspicuous when observations are made from a vessel drifting with the ice, because every change in wind direction and velocity brings about a corresponding change in the drift of the ice. Careful observations of the ice drift were conducted by Brennecke during the drift of the *Deutschland* in the Weddell Sea in 1911–1912. He found that the direction of the ice drift deviated on an average 34° from the wind direction, and not 45° as required by the Ekman theory of wind currents (p. 125). The discrepancy was explained by Rossby and Montgomery as due to a layer of shearing motion in the water directly below the ice, but it may be due in part to the resistance offered by the ice because the wind does not blow uniformly over large areas, wherefore the ice is packed together in some areas and torn asunder in others. The ice resistance appears, however, to be small in the Antarctic compared to that in the Arctic, because in the Antarctic the drift of the ice is not impeded by land barriers on all sides. The antarctic pack ice consists therefore of larger ice floes than the arctic pack ice and is less broken and piled up.

The antarctic pack ice has in all months of the year a well-defined northern boundary beyond which no great amounts of scattered ice are encountered. The northern limits of the ice at the end of the winter (October) and the end of the summer (March) are shown in fig. 62. Parts of the antarctic coast are always ice-free in summer, such as the Pacific side of Graham Land, and other parts are often ice-free, such as the coasts to the south of Australia and Africa. The eastern areas of the Weddell Sea and the Ross Sea are always ice-free. The Weddell Sea can often be entered from the east without encountering pack ice, but in order to enter the Ross Sea it is always necessary to pass through a broad belt of pack ice.

The great icebergs of the Antarctic originate mainly from the *shelf ice*, which represents the direct continuation of the antarctic ice cap where it extends into the shallow waters surrounding the continent. The shelf ice is partly afloat, and when it is pushed far out it breaks off in enormous pieces that may be tens of kilometers wide and up to 100 km long and rise 90 m out of the water, corresponding to a thickness of about 800 m. These giant bergs, which have occasionally been mistaken for islands, drift through the pack ice and often move in a direction opposite to the ice, because they are carried by currents and are less influenced by wind. Since the melting of these bergs takes a long time, they drift to greater distances from the continent than the sea ice. Fig. 62 also shows the average northern boundary of icebergs according to British Admiralty chart No. 1241, but icebergs or remains of icebergs have been reported much further north than is indicated by this average limit. On April 30, 1894, a small piece of floating ice was sighted in lat. 26°30′S, long. 25°40′W, but otherwise remains of icebergs are rarely reported from localities north of 35°S in the Atlantic Ocean, north of 45°S in the Indian Ocean, and north of 50°S in the Pacific Ocean.

ICE AND ICEBERGS IN THE ARCTIC. The Polar Sea with its adjacent seas, the western portion of the Norwegian Sea, Baffin Bay, and the western portion of the Labrador Sea are covered by sea ice during the greater part of the year, whereas, under the influence of the warm Atlantic water the west coast of Norway is always ice-free except for occasional freezing over of the inner parts of fjords.

The arctic pack ice that covers most areas is more broken and piled up than the antarctic pack ice. Large, flat ice floes are rare, but fields of hummocked ice and pressure ridges rising up to 5 or 6 m above the general level of the ice are frequently found. The broken-up and rugged appearance of the arctic pack ice is ascribed to the action of the wind in connection with the restricted freedom of motion of the ice due to land barriers on all sides.

The wind drift of the ice has been discussed by Nansen and by Sverdrup, who found that the direction of the drift deviated, on an average, 28° and 33°, respectively, from the direction of the wind, instead of 45° as required by Ekman's theory of wind currents. The discrepancy is due to the resistance against motion offered by the ice itself. Because this resistance is greatest at the end of the winter, when the ice is most closely packed, the deviation of the ice drift from the wind direction is smallest at that time of the year. Similarly the wind factor—that is, the ratio between the velocity of the ice drift and the wind velocity—is smaller at the end of the winter than in summer, varying between 1.4×10^{-2} in April to 2.4×10^{-2} in September.

Under the influence of variable winds the ice, in all seasons of the year, is torn apart in some localities where lanes of open water are formed, and is packed together in others. In winter the lanes are rapidly covered by young ice which in a week or less may reach a thickness of 50 cm, but from the end of June to the middle of August no freezing takes place, because over the entire Polar Sea the air temperature remains at zero degrees or a little above zero. Where lanes are formed the ice always breaks along a jagged line, and when the ice fields move apart they also are displaced laterally. In summer such lanes remain ice-free, and when they close, because the ice is packed together, the two sides of the lane do not fit, corners meet corners, and between the corners openings of different shapes remain, many of which are several hundred meters long. In summer the arctic ice fields are therefore not continuous, but are honeycombed to such an extent that from no point can one advance as much as 10 km in any direction without striking a large opening in the ice. This characteristic makes possible the use of a submarine for exploration of the Polar Sea, as claimed by Sir Hubert Wilkins.

In winter the average thickness of the arctic pack ice is probably 3 to 4 m, and in summer 2 to 3 m, but under pressure ridges and great hummocks the thickness is greater. Hummocked ice is often found stranded where the depth is 8 to 9 m, and in exceptional localities where strong tidal currents prevail, masses of piled-up ice have been found stranded where the depth was 20 m. The ice passes through a regular annual cycle. In summer, melting takes place for two or three months, and on an average the upper 1 m of the ice melts. In winter, ice forms on the underside of the floes, but the thicker the ice is, the slower the freezing. The average thickness of the ice depends mainly upon the rapidity of melting in summer and freezing in winter, and is therefore determined by climatic factors. Near shore, river water and warm offshore winds facilitate melting and the development of navigable lanes of open water along the coasts. In recent years the USSR has been able to take advantage of such lanes along the north coast of Siberia for establishing shipping connections with the large Siberian rivers. Aerial surveys of ice conditions have preceded the operations and ice breakers have been used where necessary.

Within the greater part of the Polar Sea the ice moves mainly under the action of the winds, but it is also carried slowly by currents toward the opening between Spitsbergen and Greenland. The speed of the current increases when approaching the opening, and great masses of ice are carried swiftly south by the East Greenland Current. Since 1894 detailed information as to the extent of the sea ice in the Norwegian Sea and in the Barents Sea has been annually compiled and published by the Danish Meteorological Institute.

The icebergs in the Arctic originate from glaciers, particularly on Northern Land, Franz Josef's Land, Spitsbergen, and Greenland. No icebergs are encountered in the Polar Sea except near Northern Land, Franz Josef's Land, and Spitsbergen, because the Greenland glaciers do not terminate in the Polar Sea. The glaciers on the first three islands are small and produce only small icebergs. By far the greater number and all large icebergs originate from the Greenland glaciers and are carried south by the East Greenland and Labrador Currents. These icebergs are generally of irregular shape, because the Greenland glaciers do not form thick shelf ice comparable to that of the Antarctic, but terminate in fjords where piece after piece breaks off as the glacier advances. Some fjords are closed by shallow sills on which the bergs run aground.

Many of the icebergs that are carried south by the East Greenland Current disintegrate before they reach Cape Farewell, the southern cape of Greenland, but others are carried around the south end of Greenland and continue to the north along the west coast. They are joined by other icebergs from the West Greenland glaciers and together with these are finally carried south by the Labrador Current. The icebergs reach farthest south off the Grand Banks of Newfoundland in the months March to June or July, when they represent a serious menace to shipping. After the *Titanic* disaster in 1911 the International Ice Patrol was established in order to safeguard shipping by reporting icebergs and predicting their probable course. The ice patrol is conducted by the U. S. Coast Guard, the publications of which contain a large amount of information as to the number, distribution, and drift of icebergs in the region of the Grand Banks. Prediction of the drift has been based successfully on currents computed from the distribution of density as observed on cruises during which several lines of oceanographic stations have been occupied in about two weeks. This work of the U. S. Coast Guard is the most outstanding example of practical application of the methods for computing ocean currents (p. 111).

CHAPTER X

Interaction Between the Atmosphere and the Oceans

Character of Interaction

From the preceding discussions of the heat budget of the oceans and the relationship between the prevailing winds and the ocean currents, it is evident that the oceans must exercise a profound influence upon the climates of the world and upon the larger and smaller features of the atmospheric circulation which together determine the weather, and that, conversely, the atmosphere controls to a great extent the oceanic circulation. The interaction, however, is so complicated that it is as yet impossible to separate cause and effect.

The ocean currents are closely related to the prevailing winds, and the energy needed for maintaining the currents is derived from the stresses that the winds exert on the sea surface. However, the currents would alter the distribution of density if this distribution were not also controlled by exchange of heat with the atmosphere and by processes of evaporation, precipitation, and radiation. Thus, the approximately stationary distribution of density in the oceans represents a state of dynamic equilibrium in which changes induced by the action of the winds are balanced by changes related to the processes of heat conduction, evaporation, precipitation, and radiation, all of which are closely related to the state of the atmosphere. It might appear, therefore, as if the oceanic circulation and the distribution of temperature and salinity in the oceans are caused by the atmospheric processes, but such a conclusion would be erroneous, because the energy that maintains the atmospheric circulation is to a great extent supplied by the oceans. It will be shown that this energy supply is very localized, owing to the character of the ocean currents, and that therefore the circulation of the atmosphere, which depends upon where energy is supplied, must be influenced by the oceanic circulation. The reasoning leads to the conclusion that one cannot deal independently with the atmosphere or the oceans, but must deal with the complete system, atmosphere-oceans. This fact has been recognized in oceanography, where one gets nowhere by neglecting the relation to the atmosphere, but in meterorology it has not yet received sufficient attention.

When dealing with the entire system the processes at the boundary between atmosphere and ocean—that is, at the sea surface—must be fully understood, and in the earlier chapters emphasis has been placed on discussion of these processes. Heat exchange and evaporation have been dealt with, and the conditions that further or reduce these processes have been described. The stress of the wind has been discussed, and the relation of the ocean currents to the prevailing winds has been dealt with. As far as the atmosphere is concerned, only the very lowest layers that are in immediate contact with the sea surface have been considered, but it will now be attempted to point out some of the larger features of the atmosphere that are directly related to the distribution of land and sea.

The Oceans and the Climate

The influence of the oceans on the climate has long been recognized. The two most extreme types of climate that are related to the distribution of land and sea are the continental and the maritime. The continental type, which is found over great inland areas, is characterized by warm summers and cold winters and, therefore, by a wide range in temperature between summer and winter, and by warm days and cool nights, and thus by a wide range between day and night. Mostly, it is also characterized by little cloudiness and little precipitation. In contrast, the maritime climate, which prevails over the oceans, is characterized by a small annual variation in temperature, by cool summers and mild winters, by a very small range of temperature between day and night, and often by great cloudiness and considerable precipitation. In middle latitudes the general circulation of the atmosphere is directed from west to east, and east coasts are, therefore, greatly under the influence of winds blowing from a large continent, whereas west coasts are under the influence of onshore winds from the sea. In these latitudes, therefore, the east coasts have a more continental climate and the west coasts a more maritime climate. In equatorial regions the circulation is from east to west and the climatic conditions of the coasts are reversed, but the differences are small.

The great difference between the continental and the maritime climates of middle and high latitudes is mainly due to the fact that the solid earth, in contrast to the ocean, does not store any appreciable amount of heat. Only the very surface of the land is subjected to temperature changes, and the surface temperature is, therefore, greatly raised in summer when heat is received from the sun, and greatly reduced in winter when excessive loss of heat by radiation takes place. Similarly, the temperature is increased during the day and reduced during the night. The ocean, on the other hand, can store large amounts of heat, because in summer processes of vertical mixing distribute the heat absorbed from the sun over a relatively thick layer of water, the temperature of which will be only moderately increased. In winter the loss of heat similarly takes place from a considerable layer of water, for which reason the temperature decrease of the surface of the sea will be small. If processes of mixing took place but if horizontal ocean currents were lacking, there would still be a marked difference between the climate of the land and that of the oceans, and a difference between the climate of the coast and that of the inland areas. The climate over the oceans would be much more uniform than that over the continents, and the climate of the coasts would be milder than that over the inland. When dealing with actual conditions it must be taken into account that ocean currents exist that also tend to modify the distribution of temperature of the surface waters and to bring into one area waters that are abnormally warm and into another area

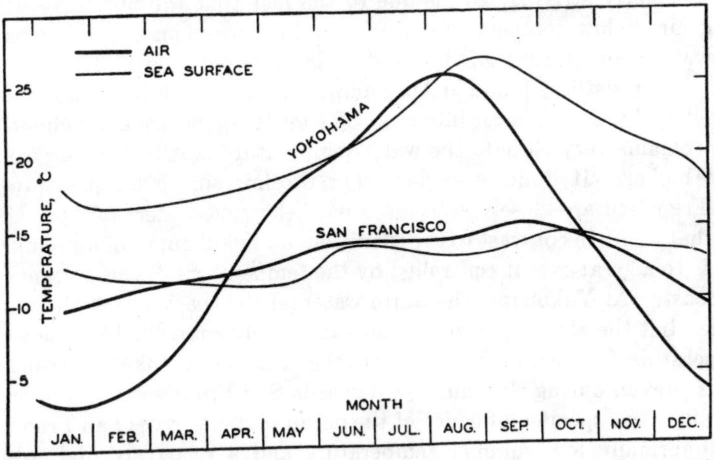

Fig. 63. Annual variation of air- and sea-surface temperatures at Yokohama and San Francisco.

waters that are abnormally cold. Thus the circulation of the ocean will alter the climate over the sea itself and will modify the climate of the coasts, because the current flowing along one coast may be abnormally warm and the current flowing along another coast may be abnormally cold. The influence will evidently depend upon the character of the currents, which, again, is controlled by the prevailing winds and the processes of heating and cooling to which the waters of these currents have been subjected in course of time.

The difference in climate between the east coasts and the west coasts in middle latitudes can be brought out by considering specific examples of difference in the annual march of temperature on both sides of the Pacific Ocean. Yokohama and San Francisco are nearly in the same latitude, but the annual march of temperature in these localities is widely different, as is evident from fig. 63, in which the mean monthly temperatures are shown. At Yokohama the average temperature of the coldest month is as low as 3.0°, but the warmest month has an average temperature of 25.4°.

The great range of 22.4° clearly demonstrates the continental character of the climate of Yokohama. In San Francisco, on the other hand, the coldest and warmest months have average temperatures of 9.7° and 15.2°, respectively, and the small range of only 5.5° clearly demonstrates that San Francisco, from this point of view, has a maritime climate.

A comparison of the air temperature with the water temperatures off the coasts is of further interest (fig. 63). Off Yokohama the water temperature varies during the year between 16.6° and 26.2°, and in nearly all months the water is warmer than the air. The great difference in winter is particularly striking, and is due to the fact that the northerly winds bring air of low temperature. Off San Francisco, on the other hand, the water temperature off the coast varies only between 11.3° and 14.6°. The coldest water is found in April and May, because in these months the upwelling (p. 201) is most intense. The air temperature throughout the year remains very close to the water temperature, but is generally somewhat higher. It is quite evident that the low spring temperatures in San Francisco are closely associated with the cold waters off the coast, and that there, in contrast to the conditions at Yokohama, the air temperature is to a great extent controlled by the temperature of the currents off the coast. At Yokohama the warm water off the coast makes the winter milder, but the air temperature is not completely controlled by the water temperature. The difference is due to the fact that in Yokohama offshore winds prevail during the winter, whereas in San Francisco onshore winds prevail. The specific influence of the ocean currents gives San Francisco an abnormally low summer temperature and a relatively high winter temperature. Any number of examples could be added, but it is sufficient to refer to the Climatological Charts of the Oceans. These charts, together with representations of the currents of the oceans, clearly demonstrate the relation between the oceanic circulation and the climates of the coasts.

The Oceans and the Weather

The relations between the ocean currents, temperatures, and the weather are equally important, but, being more difficult to examine, these relations have received little attention. For this reason we shall deal mainly with one subject—namely, the regional difference in evaporation and in heat exchange between the atmosphere and the ocean.

It was emphasized that great evaporation takes place where and when the surface temperature of the water is higher than the temperature of the air a few meters above the water (p. 64). From the discussion of the ocean currents we know that in middle latitudes the currents carry warm water away from the Equator along the eastern coasts of the continents, and cold waters toward the Equator along the western coasts. In winter the winds on an eastern coast in middle latitudes blow, in general, from the west carrying cold continental air, whereas on a western coast the winds, while also blowing from the west, carry relatively warm maritime air. Thus, in middle latitudes, the waters off the continental east coast are in winter much warmer than the air, but off the west coast the waters may be colder. Consequently, one must expect that in winter the evaporation in middle latitudes is localized, taking place primarily off the eastern coasts of the continents. The above reasoning applies equally well to the amounts of sensible heat given off from the ocean; that is, in middle latitudes, in winter, heat is conducted from the oceans to the atmosphere mainly off the east coasts of the continents. The condensation of water vapor in the atmosphere is one of the major sources of heat supply to the atmosphere as a whole (p. 5), and in middle and high latitudes is probably the major one, particularly in winter. Since evaporation is localized, it seems reasonable that condensation is also localized, so that the regional heat supply to the atmosphere depends upon the interaction between the sea and the atmosphere. Both the major features and the details of the atmospheric circulation are dependent upon where the supply of energy takes place, and it follows therefore that the weather over wide oceanic areas and on coasts and even over inland areas must reflect the results of the interaction.

The conclusions as to the localization of the regions of maximum evaporation and heat transfer have been confirmed by Jacobs for the North Atlantic and the North Pacific. It was shown (p. 61) that both evaporation and heat transfer can be computed from meteorological data and sea surface temperatures, provided that certain results in fluid mechanics are applicable to the air flow directly over the sea, and that the results of a few measurements of humidity gradients over the sea can be generalized. The evaporation, E, can be derived by the formula

$$E = k(e_w - e_a)W,$$

where e_w is the vapor pressure at the sea surface, e_a is the vapor pressure in the air, and W is the wind velocity.

The evaporation can be expressed as the thickness of the water layer which evaporates in a given length of time, such as centimeters of water per day, or in units of heat required for the evaporation from a given area in a given time, such as gram calories per square centimeter per minute or per day. If the evaporation is expressed as heat, Q_e, the corresponding amount of heat given off to the atmosphere, Q_h, is (p. 63)

$$Q_h = 0.64\, Q_e \frac{\vartheta_w - \vartheta_a}{e_w - e_a} \frac{p}{1000},$$

where ϑ_w is the surface temperature of the water, ϑ_a is the temperature of the air, and p is the atmospheric pressure. The total energy lost by the water is therefore

$$Q_a = Q_e + Q_h = k(e_w - e_a)\left(1 + 0.64\,\frac{\vartheta_w - \vartheta_a}{e_w - e_a}\,\frac{p}{1000}\right)W.$$

The numerical values of the constant k depend upon the units employed and upon the heights above the sea surface at which the vapor pressure in the air and the wind velocity are measured. The constant can be determined on a semi-theoretical basis, or, when average values of observed differences and wind velocities are introduced, it can also be derived by requiring that the computed annual energy loss from the the oceans must equal the total radiation surplus (p. 67).

The latter method was adopted by Jacobs, who based his investigations on the climatological charts of the oceans, published by the U. S. Weather Bureau. These charts contain the average wind velocities in the four seasons for all oceans and for the North Atlantic and the North Pacific; they also contain charts for seasons of the temperature difference $(\vartheta_w - \vartheta_a)$, for values of the sea surface temperature, and for the wet-bulb depression. Thus, the necessary data are available for carrying out the computations for the two oceans.

Figs. 64 and 65 show Jacobs' charts for the total amount of energy given off from the sea surface in summer and winter, expressed in g cal/cm²/day. The charts clearly demonstrate that in the Tropics nearly the same amounts of heat are given off in the two seasons, whereas in middle and higher latitudes the atmosphere receives very little energy from the oceans in summer but great quantities in winter. It is also seen that in winter by far the greater amounts of energy are transferred to the air off the east coasts of the continents, in the Atlantic Ocean from the Gulf Stream system, and in the Pacific Ocean from the Kuroshio system.

The importance of the ocean currents to the localization of the areas in which energy is given off to the atmosphere is further illustrated by figs. 66 and 67. Fig. 66 shows the net *annual* surplus of radiation that the water receives. It has been computed by correcting Kimball's charts of the incoming radiation from the sun and sky for reflection loss at the sea surface and by subtracting the net back radiation to the atmosphere as derived from the sea surface temperature, the humidity of the air over the oceans, and the cloudiness (p. 59). During the year the oceans in all latitudes receive a surplus of radiation, but north of latitude 25°N the surplus decreases with increasing distance from the Equator. In latitudes 10°N to 45°N it is smaller off the east coast of the continents than off the west coasts because of the difference in cloudiness.

If the oceans were at rest and if the mean annual temperature remained constant, the annual radiation surplus in every locality would have to be given off to the atmosphere—that is, in every locality one would have $\bar{Q}_r = \bar{Q}_a$. In the presence of currents this relation is valid

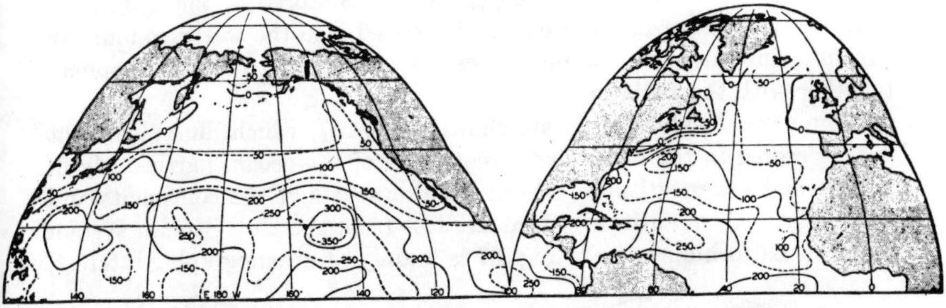

Fig. 64. Total amount of energy given off in summer from the North Pacific and North Atlantic Oceans (gram calories per square centimeter per day).

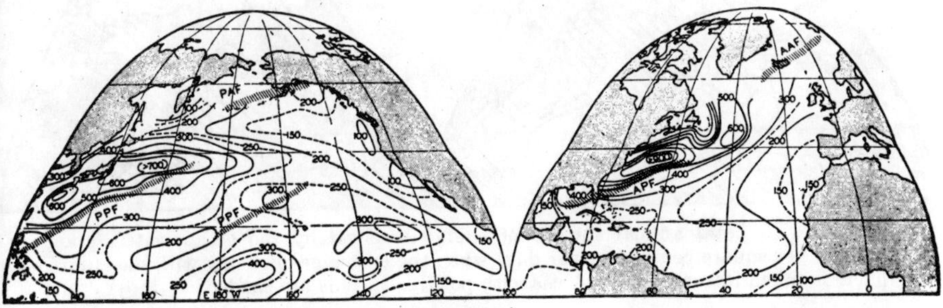

Fig. 65. Total amount of energy given off in winter from the North Pacific and North Atlantic Oceans (gram calories per square centimeter per day). Location of polar and arctic fronts is indicated.

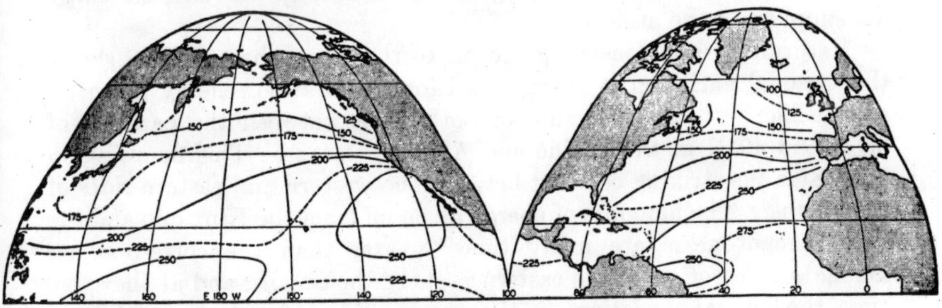

Fig. 66. Net annual surplus of radiation penetrating the water surface (gram calories per square centimeter per day).

for the oceans as a whole, but for any given locality

$$\bar{Q}_v = \bar{Q}_r - \bar{Q}_a.$$

Here $\bar{Q}_v$ represents the total surplus energy received by every square centimeter of the surface, taking the exchange of energy with the atmosphere into account. This surplus must, if positive, be carried out of the water column by currents and processes of mixing, and it then represents the part of the radiation surplus, $\bar{Q}_r$, which is stored in the water. A negative surplus means that energy is carried into the water column by currents and processes of mixing and is given off to the atmosphere together with the radiation surplus.

The annual values of $\bar{Q}_v$ are shown in fig. 67, which illustrates that heat is stored in the water over great areas of the ocean, particularly in middle and lower latitudes off the western coasts of the continents and that enormous quantities of heat are lost from the Gulf Stream system and the Kuroshio system. Fig. 67 is mainly of oceanographic interest,

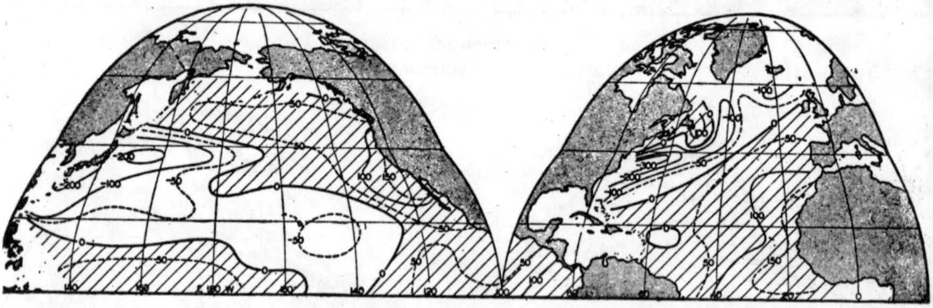

Fig. 67. Total annual surplus of energy received by the ocean water (gram calories per square centimeter per day) when the exchange of energy with the atmosphere is taken into account. Areas with positive surplus are shaded.

but it serves in this connection to emphasize the part played by the ocean currents in *concentrating* the radiation surplus that the oceans receive and thus localizing the regions in which large amounts of energy are supplied to the atmosphere.

Returning to the energy given off to the atmosphere, fig. 68 shows the seasonal variation in energy used for evaporation, energy given off as sensible heat, and the sum of both items between the parallels of 35°N and 40°N in the Pacific and Atlantic Oceans. This figure again illustrates the striking contrast between the western and eastern sides of the oceans. The amounts of energy given off from the Kuroshio and the Gulf Stream areas are many times greater than the corresponding amounts given off from the eastern parts of the oceans, and at the same time the annual variation on the western side is both absolutely and relatively (that is, expressed in percentage of the mean annual amounts) many times greater than on the eastern side. The sharp peak of the energy curve in the Gulf Stream as contrasted to the broader band over the Kuroshio region is related to the course of the two currents. Between latitudes 35° and 40° N the Gulf Stream flows to the east-northeast, whereas the Kuroshio continues nearly due east. The fact that only small amounts of energy are given off between longitude 160°W and the west coast of the United States clearly demonstrates that the "warm Japan Current" does not exercise any direct influence on the climate of

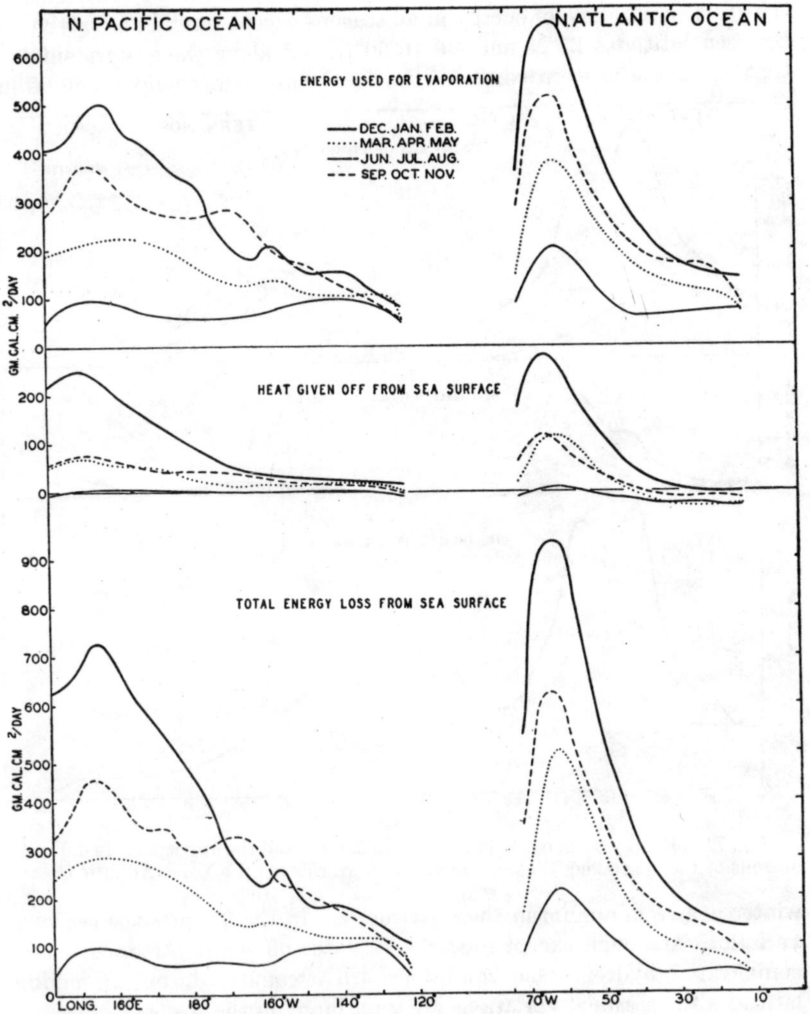

Fig. 68. Energy exchange between the sea surface and the atmosphere in different seasons of the year between the parallels of 35°N and 40°N in the Pacific and Atlantic Oceans.

the North American west coast, because it does not reach within several hundred miles of that coast (p. 200).

The contrasts between the two sides of the oceans are again brought out in figs. 69 and 70, which show the variation with latitude of energy given off in different seasons along the western and eastern sides of the two oceans at distances of a few hundred kilometers from the coasts. Along the western sides—that is, off the eastern coasts of the continents—the greatest amounts of energy in all seasons except summer are given off between latitudes 25°N and 40° to 50°N, but along the eastern sides—that is, off the western coasts of the continents—the energy given off in

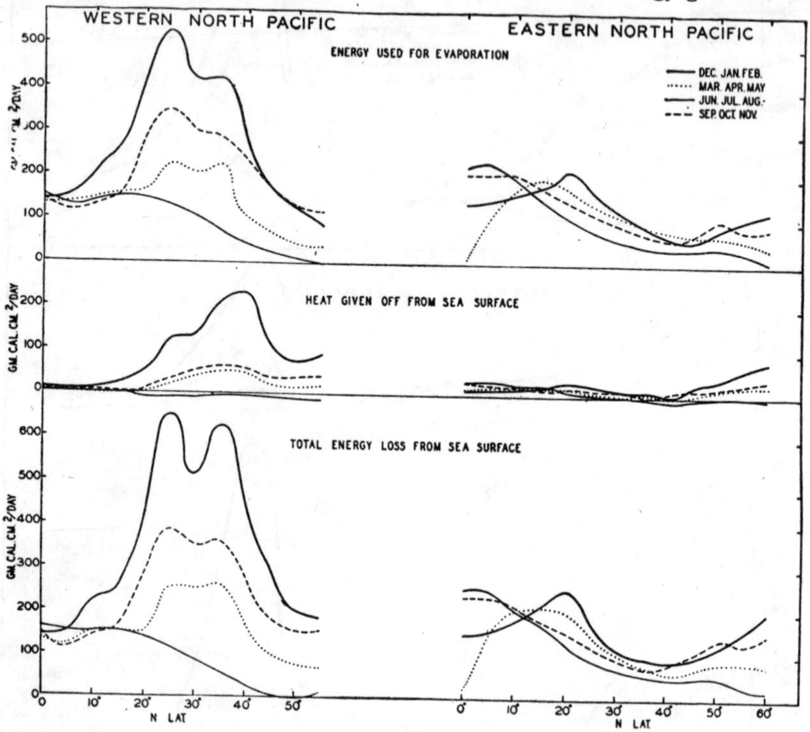

Fig. 70. Energy exchange between the sea surface and the atmosphere in different seasons of the year along the western and eastern sides of the North Atlantic Ocean.

winter is at a minimum in these latitudes. In the Tropics the seasonal variations are small except near the Equator off South America, where complicated hydrographic conditions are encountered, but in middle latitudes the seasonal variations are very large on the western sides.

The meteorological consequences of these conditions have not yet been examined, and it is therefore impossible to indicate more than a few. In fig. 64 the average locations in winter of the Polar and Arctic Fronts over the Pacific and the Atlantic are indicated according to Petterssen. The two main fronts both lie on the eastern side of the areas from which large amounts of energy are supplied to the atmosphere. The secondary Polar Front in the Pacific, which Petterssen places to the west of the Hawaiian Islands, also lies close to a region that shows a secondary maximum of energy supply. In the Pacific the Arctic Front crosses the Aleutian Islands and passes over the eastern part of Bering Sea and the

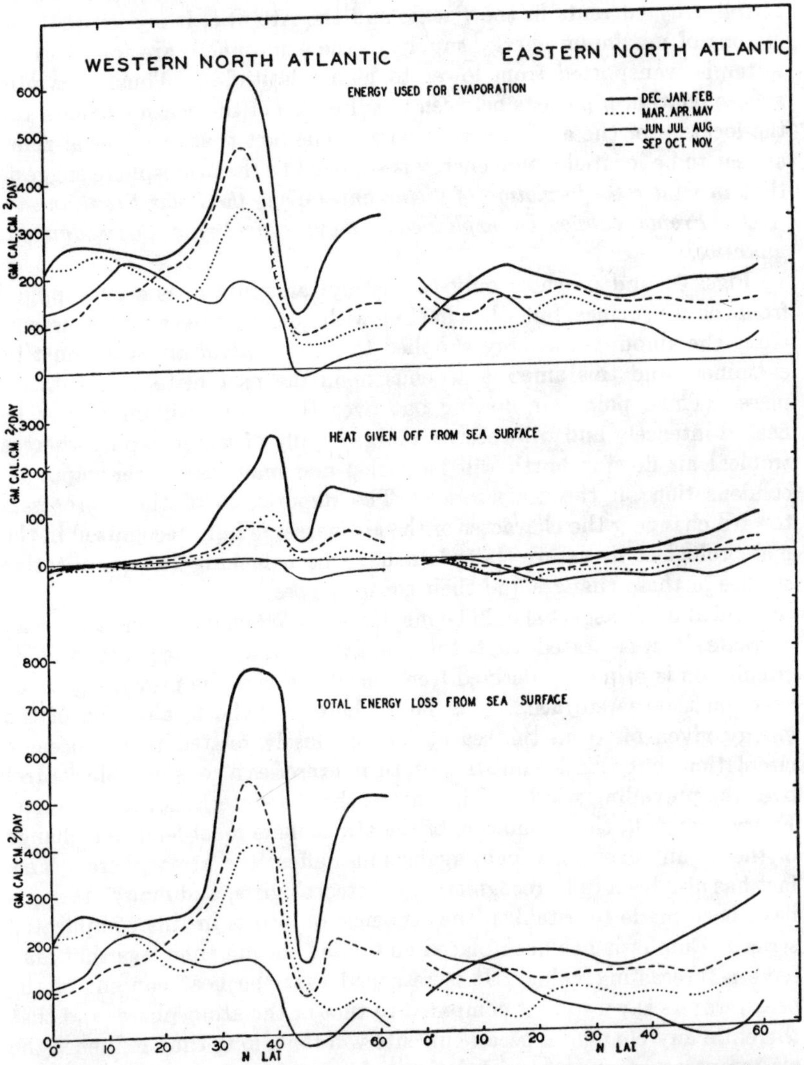

Fig. 69. Energy exchange between the sea surface and the atmosphere in different seasons of the year along the western and eastern sides of the North Pacific Ocean.

Gulf of Alaska, where the energy supply reaches another secondary maximum. In the Atlantic the Arctic Front crosses Iceland. The isolines for energy supply do not extend beyond Iceland, but it is evident that the Atlantic Arctic Front also is located in a region of great energy supply.

By comparing fig. 64 with figs. 56 and 42, which show the transport of water by currents in the Pacific and the Atlantic, it is seen that the regions of maximum energy supply to the atmosphere are found where water is transported from lower to higher latitudes. Thus, in winter a close relationship exists between the direction of the ocean currents and the location of the atmospheric fronts. The fact that the frontal zones appear to be located where energy is supplied to the atmosphere suggests that *in winter the formation of disturbances along the Polar Front and the Arctic Fronts cannot be explained without considering thermodynamic processes.*

Figs. 64 and 65 show only the *average* amounts of energy supplied from the sea surface, but when dealing with questions in synoptic meteorology the amount of energy supplied to an *individual air mass* must be examined, and this amount depends upon the past history of that air mass. Thus, polar air flowing out over the ocean will in winter be heated intensely and will receive a large supply of water vapor, whereas tropical air flowing north will be cooled and may lose water vapor by condensation on the sea surface. The importance of these processes toward changing the character of the air masses is fully recognized in the meteorological literature, but data have been lacking for quantitative studies of these changes and their consequences.

Still another aspect should be mentioned. When discussing the ocean currents it was stated that the energy for maintaining the oceanic circulation is primarily derived from the stress that the prevailing winds exert on the sea surface. It is shown here that the localization of the energy given off from the sea surface is closely related to the oceanic circulation, but this localization, in turn, exercises a considerable control over the prevailing winds. This means that the interaction is complete. Every change in the circulation of the atmosphere must lead to a change of the ocean currents, which, again, must affect the atmosphere. This fact has also been fully recognized by meteorologists, and many attempts have been made to establish the sequence of events in this complicated series. Emphasis has been placed on finding certain time lags that may have a forecasting value. It is reasoned that the heat content of the ocean waters is very great compared to that of the atmosphere, and that therefore any change in ocean currents will for a long time influence the air temperature and the circulation. In some instances a close correlation has been found between variations in ocean temperature and subsequent variations in air temperature, such as those along the west coast of Norway, where, according to Helland-Hansen and Nansen, high winter temperatures are experienced if, in the preceding summer, the heat content of the Atlantic water off the coast was high, and *vice versa.*

The cause of the variations in the heat content of the water or such variations in volume transport as are described on p. 176, however, is not understood.

It is not yet possible to deal with the system atmosphere-ocean as one unit, but it is obvious that, in treating separately the circulation of the atmosphere, a thorough consideration of the interaction between the atmosphere and the oceans is necessary.

Subject Index

Index of Names